Springer Tracts in Civil Engineering

Springer Tracts in Civil Engineering (STCE) publishes the latest developments in Civil Engineering - quickly, informally and in top quality. The series scope includes monographs, professional books, graduate textbooks and edited volumes, as well as outstanding PhD theses. Its goal is to cover all the main branches of civil engineering, both theoretical and applied, including:

- Construction and Structural Mechanics
- Building Materials
- Concrete, Steel and Timber Structures
- Geotechnical Engineering
- Earthquake Engineering
- Coastal Engineering; Ocean and Offshore Engineering
- Hydraulics, Hydrology and Water Resources Engineering
- Environmental Engineering and Sustainability
- Structural Health and Monitoring
- Surveying and Geographical Information Systems
- Heating, Ventilation and Air Conditioning (HVAC)
- Transportation and Traffic
- Risk Analysis
- Safety and Security

Indexed by Scopus

To submit a proposal or request further information, please contact: Pierpaolo Riva at Pierpaolo.Riva@springer.com, or Li Shen at Li.Shen@springer.com

More information about this series at http://www.springer.com/series/15088

Marco Migliore · Cinzia Talamo ·
Giancarlo Paganin

Strategies for Circular Economy and Cross-sectoral Exchanges for Sustainable Building Products

Preventing and Recycling Waste

Springer

Marco Migliore
Department of Architecture, Built environment and Construction engineering (ABC)
Politecnico di Milano
Milan, Italy

Cinzia Talamo
Department of Architecture, Built environment and Construction engineering (ABC)
Politecnico di Milano
Milan, Italy

Giancarlo Paganin
Department of Architecture and Urban Studies (DASTU)
Politecnico di Milano
Milan, Italy

ISSN 2366-259X ISSN 2366-2603 (electronic)
Springer Tracts in Civil Engineering
ISBN 978-3-030-30320-4 ISBN 978-3-030-30318-1 (eBook)
https://doi.org/10.1007/978-3-030-30318-1

This Springer imprint is published by the registered company Springer Nature Switzerland AG
The registered company address is: Gewerbestrasse 11, 6330 Cham, Switzerland

Preface

In recent years, the issue of waste management has become a central topic for the industrialized economies worldwide and, in a broader sense, for the global community. The World Bank estimates that every year the worlds' cities generate approximately 2 billion tons of municipal solid waste (MSW), and this huge amount of waste—due to the population growth and urbanization—is expected to increase to reach 3.40 billion tons in 2050. Looking at the other side of the coin, i.e. the waste generated by industrial activities, even though there are not official worldwide statistics available, scholars estimate that for one ton of municipal solid waste from 2 to 10 tons of industrial waste are generated; as an example, the US Environmental Protection Agency (EPA) estimated that each year only in the USA approximately more than 7 billion tons of industrial solid waste are generated from industrial facilities. Within this framework the construction industry is still one of the major waste producers for reasons that range from the nature of the construction product itself, that is large and heavy, to the lower efficiency in resource use compared to other industries. As an example, the amount of waste generated in Europe by the construction sector in its various activities (construction and demolition waste or C&D waste) is estimated at about one-third of the total waste production of all economic sectors, and this quantity places it in the first place among the economic entities that produce waste in Europe.

This huge flow of matter cannot be managed in a sustainable way only with landfilling or burning and new processes and techniques should be defined in order to help countries to reduce the large impacts generated by household and industrial waste. The need to control, reduce or modify the huge and growing flows of materials generated by the production of waste has triggered innovation processes in the technical-scientific, socio-economic and political spheres. In this direction, the European Union has developed several theoretical reflections followed by a set of policies, directives and communications about the subject of waste. The vision proposed by these policies and the subsequent actions is to consider waste as a resource and not simply as an unwanted outcome of production and consumption processes and to reduce as much as possible the impact of waste on the environment during the disposal phase.

On the other hand, the "take-make-dispose" production and consumption model that has characterized the economic development in the past century established a direct proportionality between the production of waste and the exploitation of natural resources. Global extraction of primary materials has been estimated by UN at 70 billion tons in 2010, namely three times more than in year 1970, and this level is projected to reach 180 billion tons in 2050 if our production and consumption models will not change significantly. Concerning the primary material extraction, it is possible to notice that the role of the construction industry in the raw material acquisition is becoming more and more important in comparison to other industries. Data from the UN IRP Global Material Flows Database show that extraction of non-metallic minerals for the construction industry accounted for 32% of the global extraction of materials in 1970, but in 2017, the percentage raised to 46%. There are several reasons for this huge increase in material consumption from the construction industry: world population almost doubled from the 1970 and housing is a major concern in developing countries; the mobility trend needs for new and modern infrastructures; international trade leads to a significant increase in logistics hubs, etc.

The combination of these two phenomena—increasing waste production and increasing exploitation of natural resources—makes it necessary to develop new models and paradigms for waste management and for their reuse in place of the increasingly less available natural resources. One of these models is the model that proposes to "close the loop" in the above-mentioned linear take-make-dispose model giving rise to a new economic model that is better known by the term "circular economy". The circular economy model recognizes the crucial value of prevention and, at the same time, attributes to waste the potential of the virtuous generator of new markets, capable of activating new skills and entrepreneurship and to innovate production processes with new strategies of production and consumption. In this sense, the circular economy approach may also become an opportunity for developing countries considering that waste collection rates are highly dependent from the income level of the population and in less developed countries only 40% of the waste is collected. New activities and markets activated by the circular economy approach that gives value to waste may also generate new interest in collecting waste from the territory increasing the waste collection rates with positive effects for the environment and for the human health.

Starting from these premises, this book analyses the wide and complex theme of the circular economy from the perspective of strategies for the reuse/recycling of waste, assuming as an area of interest the construction sector and developing some key assumptions:

- the theme of waste reuse/recycling must be considered in a logic of cross-sectorality, recognizing the need to enhance the "dialogue" between different sectors, so that what is a waste or a by-product for a production process can be considered a secondary raw material for another production process;

- pre-consume waste appears to be particularly interesting for the recycling market for several reasons: its production over time is continuous, both the quantity produced and its chemical-physical characteristics are known and predictable; it is geographically located and immediately available;
- the construction sector, strongly involved in the production of waste, can reduce part of its environmental impacts—deriving from the extraction of materials, procurement, production and demolition processes—enhancing its capacity to use secondary raw materials and by-products from other sectors;
- the manufacturing sector, strongly involved in the production of construction materials and components, can experiment interesting forms of prevention/ recycling of pre-consumer waste and at the same time can provide by-products that can be used in other production chains according to the logic of cross-sectoral circularity;
- information has a fundamental role in recognizing and enhancing the market possibilities of by-products and in supporting the creation and connotation of recycling chains.

So declined, the topic of pre-consumer waste recycling, in a circular economy perspective, presupposes the coexistence and interaction of a plurality of subjects. In this approach, considering the possibility for the construction sector to express a demand for secondary raw materials, the book is addressed to:

- the producers of building materials and components that can become the expression of an offer of recyclable waste for other sectors;
- the producers of building materials and components that can become an expression of the demand for recyclable waste from other sectors;
- the building designers that can push and connote this demand and this offer and can guide environmental quality strategies;
- the environmental certifiers that can support this supply and demand by highlighting the advantages of products involved in recycling processes and by expressing an information request (in terms of data needed) with regard to the environmental parameters included in the assessment procedures.

The book is divided into nine chapters:

In Chap. 1, Cinzia Talamo gives an overview of the EU waste policy in order to outline the integrated strategies that orient the regulatory and procedural instruments, develop the infrastructure measures, finance the research, boost the market of secondary raw materials, support the improvement of production models and regulates the information flows.

In Chap. 2, Cinzia Talamo discusses the theme of construction and demolition waste and the role of recycling and reuse in reducing costs and negative environmental impacts, related to the extraction, processing and production of construction materials. The chapter analyses the characteristics of the CDW and investigates the key actions for reducing the general environmental impacts of the construction sector related to waste production, focusing on barriers and drivers for the development of a market of secondary materials and building components.

In Chap. 3, Cinzia Talamo discusses the reuse, considered in the perspective of waste prevention and analyses the MS approaches towards reuse and highlights the barriers and drivers that may influence its development as a widespread practice.

In Chap. 4, Marco Migliore presents the financing systems activated by the European Union, in order to promote and stimulate forms of innovation aimed at the preservation of the environment and to practices of the circular economy. In particular, the chapter examines some significant projects (especially related to the building sector, oriented to enhance and reconsider scraps and waste as a secondary raw material) and highlights the main trends through some interpretation keys.

In Chap. 5, Marco Migliore describes the results of a research carried out with the contribution of *Fondazione Fratelli Confalonieri di Milano* on the subject of the cross-sectoral valorization of waste in the building sector. The contribution highlights the role and the importance that information and its standardization can have in a circular economy scenario aimed at the enhancement of pre-consumer waste.

In Chap. 6, Giancarlo Paganin discusses the reuse of scraps from agricultural activities to the construction industry presenting some of the actual and future trends in the research that takes into account the cross-sectoral stream of matter from agriculture and livestock to the construction industry.

In Chap. 7, Giancarlo Paganin presents a review of the different systems for the attestation of conformity of materials and products including a determined percentage of recycled materials.

In Chap. 8, Nazly Atta introduces the topic of Smart Waste Management—enabled by the adoption of the latest innovations in the field of Information and Communication Technology (ICT)—discussing the concept of Information Platform as useful support tool for the establishment of collaborative networks for industrial symbiosis.

In Chap. 9, Marcella Bonanomi discusses a possibility to couple two topical subjects—one is the integrated project delivery and the other deals with participatory organization or Living Lab approach—as a strategy to foster and facilitate smart waste networks in the context of the building sector.

Milan, Italy

Marco Migliore
Cinzia Talamo
Giancarlo Paganin

Contents

Chapter 1
Waste and Circular Economy in the European Policies

Abstract The present chapter deals with the EU waste policies in order to: outline the integrated strategies that orient the regulatory and procedural instruments, develop the infrastructure measures, finance the research, boost the market of secondary raw materials, support the improvement of production models, regulate the information flows. In order to refer the subject of waste to the wider field of the circular economy, various EU waste initiatives, communications from the Commission to the European Parliament, programmes and laws are investigated, highlighting the essential features of the evolving scenario. In particular, the Waste Framework Directive 2008/98/EC, considered as the fundamental legal framework for treating waste in the EU, is analyzed in order to investigate the possible strategies for the application of the waste hierarchy to the construction sector in the perspective of the circular economy.

Keywords Prevention and recycling of waste · Circular economy · Construction sector

1.1 The Waste Issues in the EU Environmental Policies

Within the scenario of the European environmental policies, the issue of waste from manufacturing sectors and from construction and demolition activities is becoming more and more important, involving approaches and practices related to eco-innovation [1–4], industrial ecology [5, 6] and industrial symbiosis [7, 8].

Prevention and recycling of waste represent the long-term goals of the EU waste policy, oriented to increase the EU's resource-efficiency and reduce the negative environmental and health impacts over the life-cycle of resources.

The adoption in 2005 of the Thematic Strategy on the Prevention and Recycling of Waste, described in the COM (2005) 666 [9], is emblematic of the approach of

This chapter is authored by Cinzia Talamo, Politecnico di Milano, Department of Architecture, Built Environment and Construction Engineering, Milan, Italy. e-mail: cinzia.talamo@polimi.it.

M. Migliore et al., *Strategies for Circular Economy and Cross-sectoral Exchanges for Sustainable Building Products*, Springer Tracts in Civil Engineering,
https://doi.org/10.1007/978-3-030-30318-1_1

the European Commission in promoting waste prevention, reuse and recycling, with waste disposal only as last expedient [10].[1]

Actually, a first step taken by the Commission was to adopt a consultation document in 2003, entitled "Towards a thematic strategy on the prevention and recycling of waste" [11]. This preliminary communication considered the current situation and the most significant trends and proposed options for the way forward. It also highlighted unsustainable developments and areas where change was needed, but remained open as to the way forward. It launched a wide debate on a range of topics, including key issues, such as the definition of waste and the waste hierarchy.[2]

In particular, the Thematic Strategy on the Prevention and Recycling of Waste assumed the EU long-term goal to become an economically and environmentally efficient recycling society, that seeks to avoid waste and uses waste as a resource also with the support of environmental reference standards, able to facilitate the internal market in the recycling and recovery activities. The Thematic Strategy on the Prevention and Recycling of Waste represented an initial step towards adapting a regulatory Framework in terms of waste. It identified some key actions, that oriented many of the strategies and legislative acts, developed in the years to come, such as: full implementation of existing legislation, simplification and modernization of existing legislation, introduction of life-cycle thinking into waste policy, promotion of more ambitious waste prevention policies, improvement of shared knowledge and information, development of common reference standards for recycling.

The Thematic Strategy on the Prevention and Recycling has to be considered in relation to the seven thematic strategies,[3] programmed by the 6th Environment Action

---

[1] "*Since the adoption of the Strategy, the closure of sub-standard landfills and incinerators has led to significant reduction of water, soil and air pollution. A high number of sub-standard landfills have been closed down (around 3.300 closures between 2004 and 2006). Nevertheless, approximately 1000 substandard landfills have been identified by the Commission as sub-standard landfills to be upgraded or closed as soon as possible*" [10].

[2] For an overview of the evolution of the European environment waste policy see: EU Waste Policy—The Story Behind the Strategy, EC Report [12].

[3] The 6th EAP introduced the concept of Thematic Strategies. The Thematic Strategies can be seen as key elements of the Commission's Better Regulation strategy: they are all accompanied by a thorough impact assessment, assessing the economic, social and environmental impacts of different policy options. The 6th EAP Thematic Strategies covered the following fields:

- Air
- Waste prevention and recycling
- Marine Environment
- Soil
- Pesticides
- Natural resources
- Urban Environment.

See: http://ec.europa.eu/environment/archives/action-programme/strategies_en.htm.

Programme[4] (EAP6) of the European Community[5] (2002–2012). Starting from the seventies, the EAPs have traced the EU environmental policies, developing medium-term plans, providing guides, frames, priorities, principles and programmes, paving the way for initiatives and legislative proposals [14–17]. In Table 1.1 an overview of the evolution of the EU environmental policies within the EAPs.

The program of the 6th EAP,[6] notwithstanding some recognised limits,[7] in its overall structure, had the merit of contributing to the creation of a global context for the environmental policy and, with regard to waste, to the introduction of some objectives of strategic interests, such as, among others:

- the development of tools to analyse the flow of raw materials in order to evaluate the circulation (imports and exports) of raw materials and waste in the Community;
- the development and implementation of measures for the prevention and management of waste through various forms (quantitative and qualitative targets for waste reduction, encouragement for the design of environmentally-friendly products, public awareness, waste prevention, etc.);
- the development of integrated waste recycling strategies (source separation, acceptance and recycling of priority waste streams, producer responsibility, development of recycling technologies);
- specifications for the distinction between what is waste and what is not.

The subsequent 7th Environment Action Programme[8] (EAP7) faces, with a higher level of awareness,[9] the subject of waste [15]. The programme identifies three priority areas, related to actions: the first action area concerns natural capital, that is protect nature and strengthen ecological resilience; the second action area concerns the

[4]Decision No 1600/2002/EC [13] of the European Parliament and of the Council of 22 July 2002 laying down the Sixth Community Environment Action Programme.

[5]In particular, the 6th program gave an important role to the subject of waste, including it among the four main environmental priorities:

- climate change;
- nature and biodiversity;
- environment and health and quality of life;
- natural resources and waste.

[6]European Commission, Being wise with waste: the EU's approach to waste management [18].

[7]For an overview of the assessments of the 6th EAP see also:

- COM (2011) 531 [19], Communication from the Commission to the European Parliament, the Council, the European Economic and Social Committee and the Committee of the Regions. The Sixth Community Environment Action Programme, Final Assessment;
- Decision No 1386/2013/EU [20] of the European Parliament and of the Council of 20 November 2013 on a General Union Environment Action Programme to 2020 'Living well, within the limits of our planet'.

[8]General Union Environment Action Programme to 2020: http://ec.europa.eu/environment/newprg/index.htm.

[9]See Endl and Berger [15], The 7th Environment Action Programme: Reflections on sustainable development and environmental policy integration, ESDN Quarterly Report No. 32, cap.2.5.

Table 1.1 The evolution of EU environmental policies within the EAPs

EAP	Period	Dominant storylines	Empowered subject positions	Key discourse(s)
EAP1	1973–1976	Greater research determines 'the facts' and standardises measurement of pollutants Education will teach population to accept environmental responsibility	The expert	*Knowledge speaks truth to power*
EAP2	1977–1981	There are natural limits to resources and to economic activities Environmental policy constrains economic development but does not conflict with it	The expert	*Knowledge speaks truth to power* and *limits to growth*
EAP3	1982–1986	Environment provides both limits to and possibilities for economic activity Environmental policy can stimulate employment and technological innovation, to aid economic competitiveness	The public	*Limits to growth* and *ecological modernisation*
EAP4	1987–1992	Environmental policy is an economic opportunity Economic instruments can be used to ensure that 'the polluter pays'	The public, the consumer	*Ecological modernisation*
EAP5	1993–2000	Economic growth can continue as long as it respects environmental conditions The shift to sustainability enhances competitiveness and innovation	The consumer, the public, the tourist	*Sustainable development* and *ecological modernisation*

(continued)

Table 1.1 (continued)

EAP	Period	Dominant storylines	Empowered subject positions	Key discourse(s)
EAP6	2002–2012	Economic growth and environmental damage can be 'decoupled' Sustainable development offers business opportunities	The citizen, the business, the consumer	*Ecological modernisation* and *sustainable development*
EAP7	2012–2020	The 'Green Economy' can absolutely decouple economic growth and environmental degradation	The business, the citizen, the stakeholder	*Ecological modernisation*

Source Machin [17]

conditions that will help to transform the EU into a resource-efficient, low-carbon economy; the third area deals with strategies and actions aiming to reduce threats to human health and wellbeing linked to pollution, chemical substances, and the impacts of climate change.

In particular the 7th EAP:

- defines, according to the Resource Efficiency Roadmap, zero residual waste (the waste that is not prevented, reused or recycled) as a legally binding target by 2020;
- establishes prevention and preparation for reuse targets for all Member States by 2020 as well as increased recycling targets, based on the best-performing EU Member States;
- aims at EU-wide incineration and landfill ban by 2020 and at stopping funding landfills and incinerators.

1.2 From Waste to Circular Economy

The EAPs, and other initiatives, demonstrate how in the last decade the European Commission has been continuously concentrating its efforts in order to transform Europe's economy into a more sustainable one. Within this vision, the issue of waste plays an increasingly important role, especially if it is considered in the perspective of the circular economy, rising from a broader scenario of policies and strategies for the environment, closely linked to objectives concerning, at the same time, economic and social growth, innovation of production processes and consumption patterns, considered in a life cycle approach.[10]

The COM [21] 398 clearly highlighted how, in the perspective of the circular economy, the issue of waste is the link between economic, environmental and productivity goals:

"*Circular economy systems keep the added value in products for as long as possible and eliminates waste. They keep resources within the economy when a product has reached the end of its life, so that they can be productively used again and again and hence create further value. Transition to a more circular economy requires changes throughout value chains, from product design to new business and market models, from new ways of turning waste into a resource to new modes of consumer behaviour. This implies full systemic change, and innovation not only in technologies, but also in organisation, society, finance methods and policies. Even in a highly circular economy there will remain some element of linearity as virgin resources are required and residual waste is disposed of*".

Circular economy means dealing with a system-wide perspective, based on the aim of keeping products and materials at their highest value as long as possible by acting on recycle, remanufacture, and reuse. The specificity of the circular economy approach is to pursue not only environmental goals, in term of lowering current carbon dioxide emissions levels, but also to meet economically attractive opportunities, that is increase resource productivity, decrease resources dependence and waste, and increase employment and growth. Statistics (Figs. 1.1 and 1.2) and various studies[11]

---

[10]More then ten years ago, the COM [9] 666 had already highlighted the importance of a life cycle approach when dealing with the issue of waste: "*Environmental policy traditionally focused on the early and the final phases of the life cycle: extraction, processing and manufacturing at one end and waste management at the other. It is now recognised that the environmental impact of many resources is often linked to the use phase. All phases in a resource's life cycle need to be taken into account as there can be trade-offs between different phases and measures adopted to reduce environmental impact in one phase can increase the impact in another. Clearly, environmental policy needs to ensure that negative environmental impact is minimised throughout the entire life cycle of resources. By applying the life-cycle approach, priorities can be identified more easily and policies can be targeted more effectively so that the maximum benefit for the environment is achieved relative to the effort expended*".

[11]The European Innovation Partnership on Raw Materials, Raw Materials Scoreboard, European Union, 2016, reports an analysis of the values and volumes of the imports and exports of selected key waste materials traded across EU borders in 2011 (Fig. 1.3). It highlights that waste exports were significantly higher than waste imports. A number of factors drives this increase, such as: high prices for secondary materials in combination with low transportation costs, growing external demand for materials, especially from Asia, unbalanced distribution of recycling capacity among EU Member States, and the recycling policies and targets set in EU waste directives.

COM (2018) 29 [28] affirms that: "*on average, recycled materials only satisfy around 10% of the EU demand for materials, in spite of a steady improvement since 2004. For a number of bulk materials, secondary raw materials satisfy over 30% of total demand for materials (e.g. copper and nickel). However, for a large number of materials, including almost all critical raw materials, the contribution of recycled materials to satisfying the demand for raw materials is still small to negligible. This may be because it is not profitable to recycle them, the technologies to recycle them are lacking, or the materials are embedded in products kept in use for a long time (e.g. rare earth elements used in wind turbines).*".

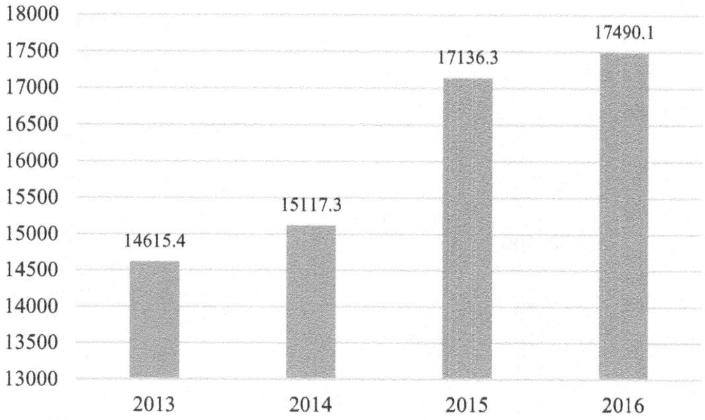

Fig. 1.1 Gross investment in tangible goods (Gross investment in tangible goods is defined as investment during the reference year in all tangible goods. Included are new and existing tangible capital goods, whether bought from third parties or produced for own use (i.e. capitalised production of tangible capital goods), having a useful life of more than one year including non-produced tangible goods such as land. Investments in intangible and financial assets are excluded.) (Mio euro) in EU (28 countries) in the following three sectors: the recycling sector, repair and reuse sector and rental and leasing sector. The recycling, repair and reuse and rental and leasing sectors are defined and approximated in terms of economic activity branches of the NACE Rev. 2 classification. *Source* Eurostat (https://ec.europa.eu/eurostat/tgm/table.do?tab=table&init=1&language=en&pcode=cei_srm030&plugin=1)

(Fig. 1.3) highlight and estimate the economic opportunities linked to circular economy [22–24]. For instance, the study, developed by Ellen MacArthur Foundation, McKinsey Centre for Business and SUN,[12] estimates that Europe will grow resource productivity by up to 3% annually, with the possibility to generate a primary resource benefit of as much as €0.6 trillion per year by 2030 to Europe's economies and €1.2 trillion in non-resource and externality benefits, bringing the annual total benefits to around €1.8 trillion versus today. This study predicts a GDP increase of as much as 7% points relative to the current development scenario, with additional positive impacts on employment.

As the study highlights, these trends make clear that the transition towards circular economy implies new models in production and consumptions requiring integrated approaches to products, services and waste: "*The smart rebound of the European economy will require game-changing strategies, breaking the paradigms prevailing since the industrial revolution. A priority is to go beyond the linear economy, where stakeholders are in traditional silos. In addition to preserving natural resources, shifting to a circular economy offers an opportunity to create new sources of wealth. The emergence of innovative models leads to collaborative dynamics across industries, cities, and communities that reveal new fields of sustainable value creation,*

[12] *Growth within: a circular economy vision for a competitive Europe*, report by the Ellen MacArthur Foundation, the McKinsey Centre for Business and Environment and the Stiftungsfonds für Umweltökonomie und Nachhaltigkeit (SUN), June 2015.

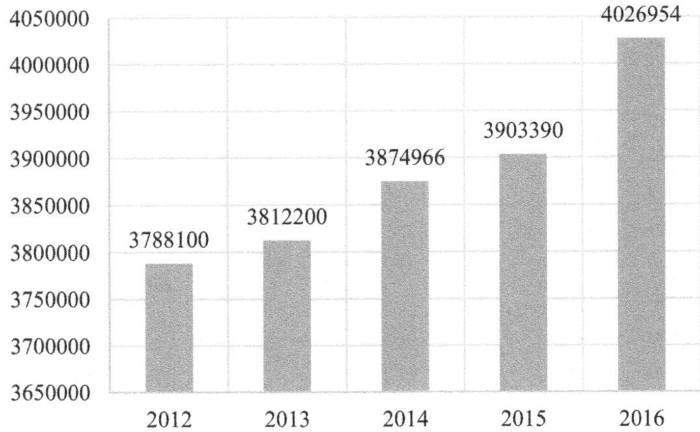

Fig. 1.2 Number of people employed (Jobs are expressed in number of persons employed and as a percentage of total employment. Number of persons employed is defined as the total number of persons who work in the observation unit, as well as persons who work outside the unit who belong to it and are paid by it. It excludes manpower supplied to the unit by other enterprises, persons carrying out repair and maintenance work in the enquiry unit on behalf of other enterprises, as well as those on compulsory military service.) in EU (28 countries) in the following three sectors: the recycling sector, repair and reuse sector and rental and leasing sector. The recycling, repair and reuse and rental and leasing sectors are defined and approximated in terms of economic activity branches of the NACE Rev. 2 classification. *Source* Eurostat

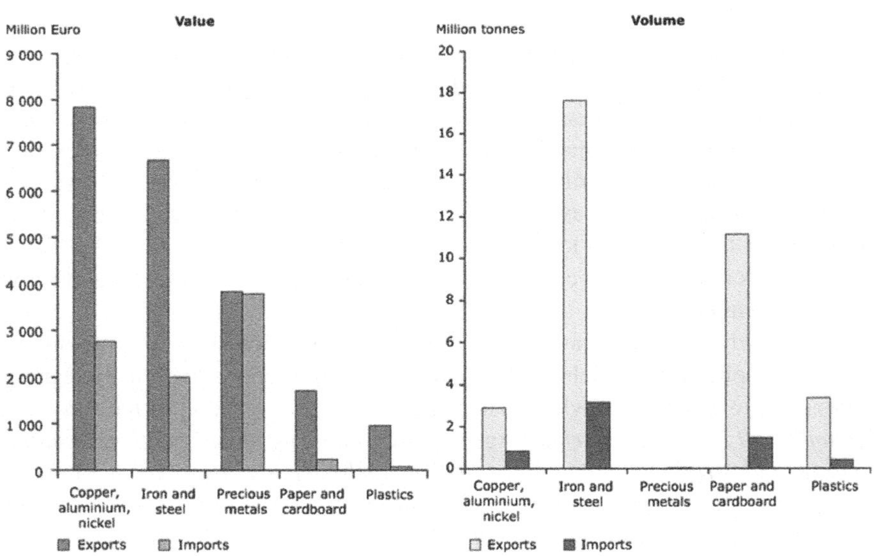

Fig. 1.3 The amounts and the value of selected non-hazardous waste materials that were exported and imported from/into the EU in 2011 (excluding Intra-EU trade). *Source* EEA (https://www.eea.europa.eu/data-and-maps/figures/trade-in-selected-waste-materials.)

such as selling services instead of products, recovering resources from waste, sharing assets, and producing green supplies. Europe offers the perfect ground for a circular economy to truly take shape and for launching disruptive models. It represents a unique opportunity but will require true vision and leadership." Source: The Ellen MacArthur Foundation, the McKinsey Centre for Business and Environment and the SUN [25].

According to the COM (2014) 398 [26], the expected transition from a linear to a circular model implies the development of innovations throughout the value chain, pursuing actions dealing with the design of production processes, products and services and at the same time creating new business opportunities [27] and supporting new markets.

COM (2018) 29 [28] reports some representative data of the transition: "*in 2014, private investments in a subset of economic sectors relevant to the circular economy are estimated to have been around EUR 15 billion in the EU (i.e. 0.1% GDP). The same year there were more than 3.9 million jobs in these sectors, an increase of 2.3% compared to 2012. In spite of the economic and financial crisis, these circular economy sectors created around EUR 141 billion of value added in 2014, which represents an increase of 6.1% compared to 2012. Several EU funding programmes are available to support the transition to a circular economy, such as the European Fund for Strategic Investments, the European Structural and Investment Funds, Horizon 2020 and the LIFE programme. In addition, in January 2017 a Circular Economy Finance Support Platform was launched. For patents on recycling and secondary raw materials, the data show an increase of 35% between 2000 and 2013. EU patents for glass recycling represent 44% of the world total for such patents, while the EU's share is 18% for plastics and 23% for paper*".

Some keywords can describe some of the conditions for this transition:

- light weighting, i.e. reducing the quantity of materials;
- durability, i.e. lengthening useful life of products;
- efficiency, i.e. reducing the use of energy and materials in production and use phases;
- substitution, i.e. cutting the use of materials that are hazardous or difficult to be recycled;
- recyclates, i.e. creating markets (linking offer and demand) for secondary raw materials,[13] also by leveraging standards, public procurement, etc.;
- eco-design, i.e. designing and promoting products that are easier to be maintained, repaired, upgraded, remanufactured or recycled, supported by the necessary services for consumers (maintenance/repair services, etc.);
- reduction, i.e. boosting and supporting waste reduction and high-quality separation in the consume phase;
- low costs, i.e. boosting systems able to minimize the costs of recycling and reuse;
- industrial symbiosis, i.e. preventing by-products from becoming wastes;

[13] As COM (2018) 29 [28] underlines: "*in the EU, the level of demand for raw materials exceeds what could be supplied even if all waste were turned into secondary raw materials. Therefore, the supply of primary raw materials will remain necessary*".

– sharing economy, i.e. boosting the consumers' choices through renting, lending or sharing services as an alternative to owning products.

1.3 Boosting the Transition in EU: The Circular Economy Action Plan

The many goals and awareness, characterizing the circular economy approach, have resulted in the ambition to implement the Circular Economy Action Plan (CEAP), aiming at promoting a balanced mix of voluntary initiatives and regulatory actions along production, consumption, waste management and secondary raw materials. In December 2015, the European Commission formally adopted a Circular Economy Action Plan, which includes measures aiming at stimulating, in the long-term, Europe's transition towards a circular economy, boosting global competitiveness, fostering a resource-efficient and competitive economy and generating new jobs [29]. In Table 1.2 an overview of the 54 actions, characterizing the (CEAP), oriented to support the circular economy in each step of the value chain[14] [7]—including production, consumption, repair and remanufacturing, waste management and secondary raw materials—that have feed back into the economy.

Regarding some strategic areas (such as plastics, food waste, construction, critical raw materials, industrial and mining waste, consumption and public procurement), the plan aims at stimulating sustainable activities in key sectors, legislative proposals and new business opportunities, including the commitments on ecodesign, the development of innovative approaches, the funding of projects, under the umbrella of the EU's Horizon 2020 research programme and the promotion of actions.

The Circular Economy Action Plan clearly highlights the role of waste related to a life cycle approach in all the actions supporting the transition towards a circular economy: "*The transition to a more circular economy, where the value of products, materials and resources is maintained in the economy for as long as possible, and*

---

[14]According to Meyer et al. [7], resource efficiency improvements all along the value chains could reduce material inputs needs by 17–24% by 2030: "*simulations till 2030 show a high potential for resource efficiency strategies to reduce material consumption and to decouple it from economic growth. Without any border adjustments the Member States of the EU might be able to install a policy mix of recycling, taxation and information and consulting that has the potential to initiate win-win situations with rising GDP and employment and falling material requirements, especially for metals, in almost all countries. (...) absolute decoupling of economic growth from material consumption does not necessarily need a general global agreement. (...) Information and consulting programs for SME's for the abatement of material inputs reduce costs, create value added for the firms under the program, but reduce sales and value added of the producers of the materials. The total effect of this is positive and induces an economy wide rebound effect, which means a rise in GDP and employment*".

See Meyer et al. [7] "Macroeconomic modelling of sustainable development and the links between the economy and the environment". Study for the European Commission (DG Environment), http://ec.europa.eu/environment/enveco/studies_modelling/pdf/report_macroeconomic.pdf 4.

Table 1.2 Actions and timetable of the EU action plan for the Circular Economy

Actions	Timetable
Production	
Emphasis on circular economy aspects in future product requirements under the ecodesign directive	2006 onwards
Ecodesign work plan 2015–2017 and request to European standardisation organisations to develop standards on material efficiency for setting future ecodesign requirements on durability, reparability and recyclability of products	December 2015
Proposal for an implementing regulation on televisions and displays	End 2015 or beginning 2016
Examine options and actions for a more coherent policy framework of the different strands of work of EU product policy in their contribution to the circular economy	2018
Include guidance on circular economy into best available techniques reference documents (BREFs) for several industrial sectors	2016 onwards
Guidance and promotion of best practices in the mining waste management plans	2018
Establishing an open, pan-European network of technological infrastructures for SMEs to integrate advanced manufacturing technologies into their production processes	2016
Examine how to improve the efficiency and uptake of the EU eco-management and audit scheme (EMAS) and the pilot programme on environmental technology verification (ETV)	2017
Develop an improved knowledge base and support to SMEs for the substitution of hazardous substances of very high concern	2018
Consumption	
Better enforcement of existing guarantees on tangible products, accompanied by a reflection on improvements (upcoming commission proposal for online sales of goods, and fitness check of consumer legislation)	2015–2017
Action on false green claims, including updated guidance on unfair commercial practices	2016
Analysis of the possibility to propose horizontal requirements on repair information provision in the context of ecodesign	2018
REFIT of ecolabel, to be followed by actions to enhance its effectiveness	2016
Assessment of the possibility of an independent testing programme on planned obsolescence	2018
Subject to evaluation of the current ongoing pilots, explore the possible uses of the product environmental footprint to measure and communicate environmental information	2016 onwards

(continued)

Table 1.2 (continued)

Actions	Timetable
Action on green public procurement: enhanced integration of circular economy requirements, support to higher uptake including through training schemes, reinforcing its use in commission procurement and EU funds	2016 onwards
Waste management	
Revised legislative proposal on waste	Dec 2015
Improved cooperation with member states for better implementation of EU waste legislation, and combat illicit shipment of end of life vehicles	2015 onwards
Stepping up enforcement of revised waste shipment regulation	2016 onwards
Promotion of industry-led voluntary certification of treatment facilities for key waste/recyclates streams	2018 onwards
Initiative on waste to energy in the framework of the Energy Union	2016
Identification and dissemination of good practices in waste collection systems	2016 onwards
Market for secondary raw materials	
Development of quality standards for secondary raw materials (in particular for plastics)	2016 onwards
Proposal for a revised fertilisers regulation	Early 2016
Proposed legislation setting minimum requirements for reused water for irrigation and groundwater recharge	2017
Promotion of safe and cost-effective water reuse, including guidance on the integration of water reuse in water planning and management, inclusion of best practices in relevant BREFs, and support to innovation (through the European innovation partnership and horizon 2020) and investments	2016–2017
Analysis and policy options to address the interface between chemicals, products and waste legislation, including how to reduce the presence and improve the tracking of chemicals of concern in products	2017
Measures to facilitate waste shipment across the EU, including electronic data exchange (and possibly other measures)	2016 onwards
Further development of the EU raw materials information system	2016 onwards
Sectorial action	
Plastics	
Strategy on plastics in the circular economy	2017
Specific action to reduce marine litter implementing the 2030 sustainable development goals	2015 onwards

(continued)

Table 1.2 (continued)

Actions	Timetable
Food waste	
Development of a common methodology and indicators to measure food waste	2016
Stakeholders platform to examine how to achieve SDGs goals on food waste, share best practice and evaluate progress	2016
Clarify relevant EU legislation related to waste, food and feed in order to facilitate food donation and utilisation of former foodstuffs for animal feed	2016
Explore options for more effective use and understanding of date marking on food	2017
Critical raw materials	
Report on critical raw materials and the circular economy	2017
Improve exchange of information between manufacturers and recyclers on electronic products	2016 onwards
European standards for material-efficient recycling of electronic waste, waste batteries and other relevant complex end-of-life products	2016 onwards
Sharing of best practice for the recovery of critical raw materials from mining waste and landfills	2017
Construction and demolition	
Pre-demolition assessment guidelines for the construction sector	2017
Voluntary industry-wide recycling protocol for construction and demolition waste	2016
Core indicators for the assessment of the lifecycle environmental performance of a building, and incentives for their use	2017 onwards
Biomass and bio-based materials	
Guidance and dissemination of best practice on the cascading use of biomass and support to innovation in this domain through Horizon 2020	2018–2019
Ensuring coherence and synergies with the circular economy when examining the sustainability of bioenergy under the Energy Union	2016
Assessment of the contribution of the 2012 bioeconomy strategy to the circular economy and possible review	2016
Innovation and investments	
Initiative "industry 2020 and the circular economy" under Horizon 2020	October 2015
Pilot project for "innovation deals" to address possible regulatory obstacles for innovators	2016
Targeted outreach to encourage applications for funding under EFSI, and support the development of projects and investment platforms relevant to the circular economy	2016 onwards

(continued)

Table 1.2 (continued)

Actions	Timetable
Targeted outreach and communication activities to assist member states and regions for the uptake of cohesion policy funds for the circular economy	2016 onwards
Support to Member States and regions to strengthen innovation for the circular economy through smart specialisation	2016 onwards
Assessment of the possibility of launching a platform together with the EIB and national banks to support the financing of the circular economy	2016
Engagement with stakeholders in the implementation of this action plan through existing fora in key sectors	2016 onwards
Support to a range of stakeholders through actions on public-private partnerships, cooperation platforms, support to voluntary business approaches, and exchanges of best practices	2015 onwards
Monitoring	
Development of a monitoring framework for the circular economy	2017

Source COM [29] 614 ANNEX

the generation of waste minimised, is an essential contribution to the EU's efforts to develop a sustainable, low carbon, resource efficient and competitive economy. Such transition is the opportunity to transform our economy and generate new and sustainable competitive advantages for Europe" [30]

In particular, regarding waste, the CEAP focuses on some key basic principles, and related goals, that should guide the EU's supporting actions [30]:

- Product design. The Commission intends to provide incentives for a design approach oriented to make products more durable or easier to repair, upgrade or remanufacture. This means, for example, to help recyclers to disassemble products in order to recover valuable materials and components[15] and to encourage a better product design by differentiating the financial contributions paid by producers under extended producer responsibility schemes on the basis of the end-of-life costs of their products;
- Production processes. The Commission should promote the sustainable sourcing of raw materials, acting on leverages such as: partnerships and trade, development policy, best practices in strategic industrial sectors through the 'best available technique reference documents' (BREFs),[16] common understandings and legislative

[15]Circular economy aspects in product design requirements can be related to the Ecodesign Directive, the objective of which is to improve the efficiency and environmental performance of energy-related products. See Directive 2009/125/EC [31] of the European Parliament and of the council of 21 October 2009 establishing a framework for the setting of ecodesign requirements for energy-related products.

[16]BREFs or 'BAT reference documents' are a series of reference documents covering, as far as is practicable, the industrial activities listed in Annex 1 to the EU's IPPC (Integrated Pollution

proposals on waste in order to clarify rules on by-products to support industrial symbiosis practices and create a level-playing field across the EU that allows waste or by-products of one industry to become inputs for another, etc.;
- Consumption. The Commission should promote waste prevention and reuse through the exchange of information and best practices and by providing Cohesion Policy funding for projects at local and regional level, including interregional cooperation. Commission can encourage actions on Green Public Procurement (GPP) and support a greater uptake of these criteria by public authorities, and reflect on how GPP could be used more widely across the EU, in particular for products or markets that have high relevance for the circular economy. Besides, the promotion of innovative models and forms of consumption can support the development of the circular economy, e.g. sharing products or infrastructure (collaborative economy), providing services rather than products, or using IT or digital platforms;
- Waste management. "*The Commission is adopting, together with this action plan, revised legislative proposals on waste comprising in particular: long-term recycling targets for municipal waste and packaging waste, and to reduce landfill; provisions to promote greater use of economic instruments; general requirements for extended producer responsibility schemes; simplification and harmonisation of definitions and calculation methods and will step up its work with Member States to improve waste management on the ground, including to avoid overcapacities in residual waste treatment.*

The Commission will assist Member States and regions to ensure that Cohesion Policy investments in the waste sector contribute to supporting the objectives of the EU waste legislation and are guided by the EU waste hierarchy" [30];

- Markets for secondary raw materials and water reuse. The boosting of the market for secondary raw materials is considered a key condition for the development of a circular economy. The Commission should develop quality standards for secondary raw materials, where they are needed, and propose improvements to the rules on 'end-of-waste'. For the boosting of a new market it is essential to facilitate the cross-border circulation of secondary raw materials to ensure that they can be traded easily across the EU: cross-border formalities can be simplified through the use of electronic data exchange. Availability of data, indicators and tools is a necessary condition for the development of a dynamic market for Secondary Raw Materials within EU. The Raw Materials Information System[17] aims to improve the availability of data on secondary raw materials.

Prevention and Control) Directive. BREF means a document, resulting from the exchange of information organized pursuant to the art. 13 of the Directive 2010/75/EU, that regards defined activities and describes techniques, present emissions and consumption levels, techniques considered for the determination of best available techniques. BREFs provide descriptions of a range of industrial processes and for example, their respective operating conditions and emission rates. Member States are required to take these documents into account when determining best available techniques generally or in specific cases under the Directive.

[17] The Circular Economy Action Plan has identified strategic areas where resource efficiency actions should be concentrated. The Raw Materials Information System supports and complements actions

In order to assess the effects of the adoption of the CEAP and to set new priorities towards the long-term objective of circular economy, the Council of the EU[18] has highlighted how monitoring trends and patterns can represent a key action to:

- understand how the various elements of the circular economy are developing over time;
- help the identification of success factors;
- assess the effectiveness of actions in Member States.

In addition, the European Parliament[19] has called upon the Commission to develop indicators on resource efficiency in order to monitor progress towards the circular economy.

Following these requests, the Commission in 2018 has presented The EU Monitoring Framework for the Circular Economy [28]. It integrates the existing Resource Efficiency Scoreboard[20] and Raw Materials Scoreboard[21] (Fig. 1.4) and includes ten key indicators, regarding each phase of the lifecycle of products; all indicators data are constantly updated and available on a dedicated website.[22] The ten key indicators (Table 1.3) are grouped into four categories, representing the basic aspects of the circular economy:

- production and consumption;
- waste management;

---

to reuse, recycle and recover materials in priority areas, by providing information and comprehensive overviews of specific industry sectors: Electrical and Electronic Equipment, including waste management processes for reuse, recycle and recover materials from household appliances and electronic devices (Waste Electrical and Electronic Equipment, WEEE); Mobility, including waste management processes for reuse, recycle and recover materials from end-of-life vehicles (ELVs) and batteries.

The European Commission's (EC) Raw Materials Information System (RMIS) is developed by the Directorate-General (DG) Joint Research Centre (JRC) in cooperation with the DG for Internal Market, Industry, Entrepreneurship and SMEs (GROWTH). The RMIS is the Commission's reference web-based knowledge platform on non-fuel, non-agricultural raw materials from primary and secondary sources. The Raw Materials Information System and EU-wide research on raw materials flows can provide an overview of: Available data from Eurostat; Available data from Horizon 2020 projects and Member State projects; Scoreboard indicators; Tools and Methods; Future data needs.

[18]Council conclusions on the EU Action Plan for the circular economy, PRESS RELEASE 367/16 20/06/2016.

[19]European Parliament Resolution of 9 July [32] on resource efficiency: moving towards a circular economy (2014/2208(INI)).

[20]Eu Resource Efficiency Scoreboard, European Union, 2016, 2015. http://ec.europa.eu/environment/resource_efficiency/targets_indicators/scoreboard/index_en.htm.

[21]The raw materials scoreboard is an initiative of the European Innovation Partnership (EIP) on Raw Materials. Its purpose is to provide quantitative data on the EIP's general objectives and on the raw materials policy context. It presents relevant and reliable information that can be used in policymaking in a variety of areas. The Scoreboard consists of 24 indicators grouped into five thematic clusters (Fig. 1.4). See https://publications.europa.eu/en/publication-detail/-/publication/1ee65e21-9ac4-11e6-868c-01aa75ed71a1.

[22]https://ec.europa.eu/eurostat/web/circular-economy.

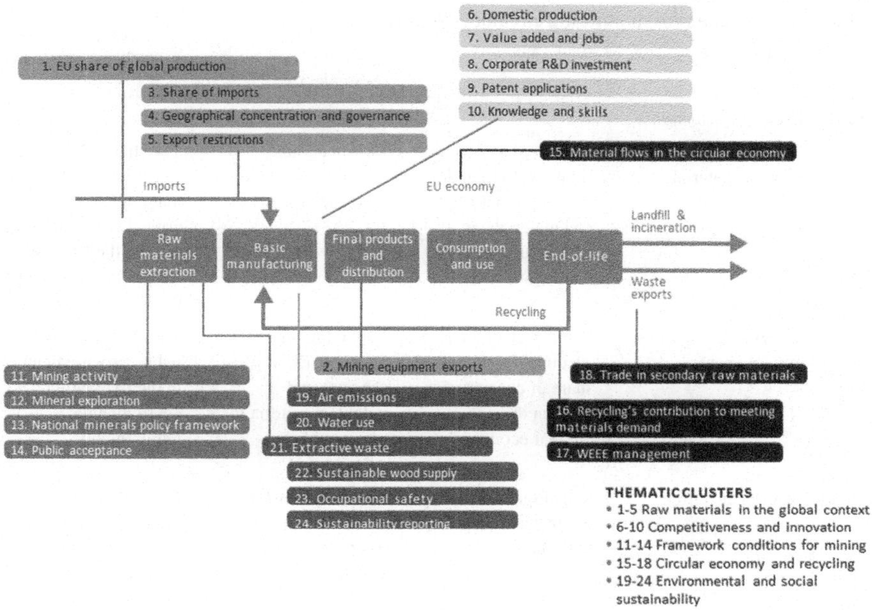

Fig. 1.4 The raw materials scoreboard thematic clusters. *Source* European Innovation Partnership [50]

– secondary raw materials;
– competitiveness and innovation.

These indicators were selected to capture the main elements of circular economy. Data availability was taken into account when choosing them, building on the Resource Efficiency Scoreboard and the Raw Materials Scoreboard. The selection of the indicators has considered existing data, various criteria (relevance, acceptance, credibility, ease of use and robustness) and the responses of public consultations. The Commission will be improving the knowledge base and data availability by developing methodologies and data collections and by funding, through Horizon 2020, several research projects that will deliver better data to complement the official statistics.

Considering the issue of information about waste, despite that Member States are required to report on waste management activities under a number of EU Directives and Regulations, the reliability of the data across the EU is highly variable, and significant problems regarding comparability across countries still remain. It is increasingly urgent to identify and analyse the main sources of uncertainties, inconsistencies and gaps in the reporting of waste statistics across the EU. Some of the main causes[23] of these problems can be summarized as follows [33]:

[23]Eunomia Research & Consulting, ENT Environment and Managemen, Ekokonsultacijos, Final Report for DG Environment under Framework Contract ENV.C.2/FRA/2013/0023, Study on Waste

Table 1.3 Indicators for monitoring circular economy

No.	Name	Relevance	EU levers (examples)	Contents
Production and consumption				
1	EU self-sufficiency for raw materials	The circular economy should help to address the supply risks for raw materials, in particular critical raw materials	Raw materials initiative; resource efficiency roadmap	The share of a selection of key materials (including critical raw materials) used in the EU that are producted within the EU
2	Green public procurement	Public procurement accounts for a large share of consumption and can drive the circular economy	Public Procurement Strategy; EU support schemes and voluntary criteria for green public procurement	The share of major public procurements in the EU that include environmental requirements
3a–c	Waste generation	In a circular economy waste generation is minimised	Waste Framework Directive; directives on specific waste streams; Strategy for Plastics	Generation of municipal waste per capita; total waste generation (excluding major mineral waste) per GDP unit and in relation to domestic material consumption
4	Food waste	Discarding food has negative environmental, climate and economic impacts	General food law regulation; waste framework directive; various initiatives (e.g. platform on food losses and food waste)	Amount of food waste generated
Waste management				
5a–b	Overall recycling rates	Increasing recycling is part of the transition to a circular economy	Waste framework directive	Recycling rate of municipal waste and of all waste except major mineral waste
6a–f	Recycling rates for specific waste streams	This reflects the progress in recycling key waste streams	Waste framework directive; landfill directive; directives on specific waste streams	Recycling rate of overall packaging waste, plastic packaging, wood packaging, waste electrical and electronic equipment, recycled biowaste per capita and recovery rate of construction and demolition waste

(continued)

Table 1.3 (continued)

No.	Name	Relevance	EU levers (examples)	Contents
Secondary raw materials				
7a–b	Contribution of recycled materials to raw materials demand	In a circular economy, secondary raw materials are commonly used to make new products	Waste framework directive; eco-design directive; EU Ecolabel; REACH; initiative on the interface between chemicals, products and waste policies; Strategy for Plastics; quality standards for secondary raw materials	Secondary raw materials' share of overall materials demand—for specific materials and for the whole economy
8	Trade in recyclable raw materials	Trade in recyclables reflects the importance of the internal market and global participation in the circular economy	Internal market policy; waste shipment regulation; trade policy	Imports and exports of selected recyclable raw materials
Competitiveness and innovation				
9a–c	Private investments, jobs and gross value added	This reflects the contribution of the circular economy to the creation of jobs and growth	Investment plan for Europe; structural and investment funds; InnovFin; circular economy finance support platform; sustainable finance strategy; green employment initiative; new skills agenda for Europe; internal market policy	Private investments, number of persons employed and gross value added in the circular economy sectors
10	Patents	Innovative technologies related to the circular economy boost the EU's global competitiveness	Horizon 2020	Number of patents related to waste management and recycling

Source COM [28] 29

- too much leeway for interpretation in legally binding obligations;
- lack of incentives for accurate data reporting;
- weakness of the approaches to addressing mis-reporting of data;
- delays in the implementation of cutting edge IT based waste data reporting systems in the Member States;

Statistics—A comprehensive review of gaps and weaknesses and key priority areas for improvement in the EU waste statistics, 2017.

– delays in the improvements of the EU-wide Specifications of definitions, criteria, targets and of standards for data verification.

Three years after the adoption, the Circular Economy Action Plan is now fully completed.

The COM (2019) 190, dealing with the implementation of the CEAP, highlights that circular economy has created new business opportunities, new business models and new markets, domestically and outside the EU, estimating that in 2016, circular activities such as repair, reuse or recycling generated almost €147 billion in value added while standing for around €17.5 billion worth of investments. Despite these encouraging estimates, the report, through Eurostat data[24] analysis, underlines that, on average, recycled materials still meet only less than 12% of the EU demand for materials (Table 1.4).

Anyway, the monitoring of the mix of voluntary initiatives and regulatory actions, activated within the Circular Economy Action Plan, demonstrates that the circular economy is now an irreversible, global trend, even if some areas not covered by the action plan should be still investigated and critical aspects should be mainly strengthen. "*The work started on chemicals, the non-toxic environment, eco-labelling and eco-innovation, critical raw materials and fertilisers needs to be accelerated if the EU want to reap the full benefit of a transition to a circular economy. Similarly consumers should be empowered to make informed choices and efforts should be enhanced by the public sector through sustainable public procuremen*t" [34]. The Staff Working Document 90 [35] provides details and references for each of the 54 actions (Table 1.2) listed in the action plan [30]. Table 1.5 highlights how the implementation of the CEAP has been continuous and characterized by some strategies. It has:

– regarded the policy, regulatory and operational frameworks;
– involved both public and private stakeholders;
– recognized information as a fundamental condition for the activation of the circular economy and has worked to the development of platforms for sharing data and experiences;
– supported the creation of a market of secondary raw materials;
– boosted voluntary protocols and certifications;
– used various EU sources of funding in order to promote and support the applied research;
– established the importance of constant monitoring.[25]

[24]https://ec.europa.eu/eurostat/tgm/table.do?tab=table&init=1&language=en&pcode=cei_srm030&plugin=1.

[25]"*The European Environment Agency shall publish, every two years, a report containing a review of the progress made in the completion and implementation of waste prevention programmes, including an assessment of the evolution as regards the prevention of waste generation for each Member State and for the Union as a whole, and as regards the decoupling of waste generation from economic growth and the transition towards a circular economy*". Directive (EU) 2018/851 [36] of the European Parliament and of the council of 30 May 2018 amending Directive 2008/98/EC on waste.

Table 1.4 Circular material use rate "*The indicator measures the share of material recovered and fed back into the economy—thus saving extraction of primary raw materials—in overall material use. The circular material use (CMU) rate is defined as the ratio of the circular use of materials to the overall material use. The overall material use is measured by summing up the aggregate domestic material consumption (DMC) and the circular use of materials. DMC is defined in economy-wide material flow accounts. The circular use of materials is approximated by the amount of waste recycled in domestic recovery plants minus imported waste destined for recovery plus exported waste destined for recovery abroad. Waste recycled in domestic recovery plants comprises the recovery operations R2 to R11—as defined in the Waste Framework Directive 75/442/EEC. The imports and exports of waste destined for recycling—i.e. the amount of imported and exported waste bound for recovery—are approximated from the European statistics on international trade in goods. A higher CMU rate value means that more secondary materials substitute for primary raw materials thus reducing the environmental impacts of extracting primary material*" (Eurostat). % of total material use

2004	2005	2006	2007	2008	2009	2010	2011	2012	2013	2014	2015	2016
8.3	8.8	9.3	9.3	9.6	10.8	11	10.6	11.3	11.5	11.4	11.4	11.7

Source Eurostat

Table 1.5 Implementation of the circular economy action plan

Actions	What has been delivered
Production	
Emphasis on circular economy aspects in future product requirements under the eco-design directive	Product requirements related to circular economy have been included in most ecodesign regulations foreseen for adoption in 2019 in accordance with to the ecodesign work plan
Ecodesign working plan 2016–2019 and request to European standardisation organisations to develop standards on material efficiency for setting future ecodesign requirements on durability, reparability and recyclability of products	– The commission adopted the ecodesign Working plan 2016–2019,[a] which proposes a list of new product groups and provides an overview of which ecodesign regulations are due for review – A set of revised ecodesign and energy labelling measures will be adopted in the first half of 2019 – The 3 European Standardisation Organisations have submitted a joint working plan, aiming at the development of standards on material efficiency aspects such as durability, reparability and recyclability
Proposal for an implementing regulation on televisions and displays	– The implementing regulation on electronic displays is part of the package of revised ecodesign and energy labelling measures proposed by the commission in 2018 and will be adopted in the first half 2019 – The new measure on electronic displays builds on the existing ecodesign measure for TVs and introduces new ecodesign requirements for televisions
Examine options and actions for a more coherent policy framework of the different strands of work of EU product policy in their contribution to the circular economy	The staff working document on sustainable products in a circular economy[b] examines the EU product policy framework. It identifies high priority product for circularity and analyses to what extent the relevant EU policies are mutually reinforcing and supporting circular economy
Include guidance on circular economy into best available techniques reference documents (BREFs) for several industrial sectors	The commission has identified key environmental issues to be addressed when revising BREFs. These include aspects on circular economy, issues concerning water use and reuse in the relevant BREFs and the overall contribution of the Industrial Emissions Directive to the circular economy
Guidance and promotion of best practices in the mining waste management plans	With a view to increase the energy efficiency of mineral extraction, thus reducing its carbon footprint, and the recovery of extractive waste by recycling and reusing, the Commission services have prepared guidance based on the best practices in the mining waste management plans
Establishing an open, pan-European network of technological infrastructures for SMEs to integrate advanced manufacturing technologies into their production processes	The EU funded project Ket4CleanProduction has established a platform[d] that gathers technology infrastructures, SME users and suppliers of innovative advanced manufacturing technologies
Examine how to improve the efficiency and uptake of the EU Eco-Management and Audit Scheme (EMAS) and the pilot programme on environmental technology verification (ETV)	The EMAS fitness check report was adopted on 30 June 2017.[e] It confirms that EMAS is a useful tool. The Commission has obtained the commitment of Member States to work for increasing the uptake of the scheme[f]

(continued)

Table 1.5 (continued)

Actions	What has been delivered
Develop an improved knowledge base and support to SMEs for the substitution of hazardous substances of very high concern	– Facilitate and disseminate best practices on the substitution of certain chemicals substances in specific areas through COSME – Support of the European resource efficiency excellence centre
Consumption	
Better enforcement of existing guarantees on tangible products, accompanied by a reflection on improvements (upcoming commission proposal for online sales of goods, and fitness check of consumer legislation)	The proposal for a Directive on the online sales of goods, presented in December 2015 was amended in 2017 to extend its scope to also cover sales of goods offline. A provisional agreement was reached by the co-legislators on 29 January 2019. The Law was finalised in May 2017. The consumer protection cooperation (CPC) Regulation, which enables consumer protection authorities to cooperate in case of trans-border infringements and to fight against widespread infringements will be applicable from 17 January 2020. As a follow-up to the fitness check, the 'new deal for consumers' package[i] was adopted by the Commission on 11 April 2018. The "new deal" and the revised CPC Regulation aim to strengthen EU consumer rights and their enforcement
Action on false green claims, including updated guidance on unfair commercial practices	Updated guidance on the Unfair Commercial Practices Directive[j] was published in May 2016
Analysis of the possibility to propose horizontal requirements on repair information provision in the context of Ecodesign	– A reparability scoring system is being developed by the Joint Research Center[k] (JRC) The JRC study will be followed up by a consumer behavioural study – The Commission is developing a scoring system on product reparability
Fitness check of Ecolabel, to be followed by actions to enhance its effectiveness	The Commission is strengthening its strategic approach to focus on product categories with significant uptake rate, and its monitoring and communication activities. The EU Ecolabel[l] catalogue has been improved. Current activities on the EU Ecolabel include the development of criteria for financial products
Assessment of the possibility of an independent testing programme on premature obsolescence	In October 2017, the Commission launched a call for an independent testing programme under H2020 to identify factors that cause premature obsolescence practices and way to address them
Subject to evaluation of the current ongoing pilots, explore the possible uses of the Product Environmental Footprint to measure and communicate environmental information	– Between 2013 and 2018, the Commission tested the application of the Product and Organisation Environmental Footprint methods on specific product groups and sectors – About 300 companies and more than 2000 stakeholders (including NGOs, public administrations, academia) worked for 5 years to test the methods.[m] A summary is presented in the Staff Working Document on Sustainable Products in a Circular Economy

(continued)

Table 1.5 (continued)

Actions	What has been delivered
Action on green public procurement: enhanced integration of circular economy requirements, support to higher uptake including through training schemes, reinforcing its use in Commission procurement and EU funds	– New/revised EU green public procurement criteria integrating circular economy requirements published since December 2015 – To support the uptake of green public procurement, the Commission published the 3rd edition of the "Buying green" handbook[n] and the brochure "Public Procurement for a Circular Economy[o] – The training toolkits are being revised and training schemes with national public authorities will take place in 2019 – A mapping exercise was initiated to identify how the uptake of green public procurement could be strengthened in the Commission's own procurement and in the spending of EU Funds – The monitoring framework for circular economy adopted in 2018 include an indicator on Green Public Procurement
Waste management	
Revised legislative proposal on waste	– The revised legislation[p] was adopted by the co-legislators on 30 May 2018 and entered into force on 4 July 2018. The outcome of the review added further ambition, in particular to ensure the application of circular economy principles to waste management – The legislative framework includes – Long-term recycling targets for municipal waste and packaging waste, and provisions reduce landfilling – Promotion of economic instruments – General requirements for extended producer responsibility schemes – Simplification and harmonisation of definitions and calculation methods – Strengthened separate collection provisions for plastic, glass, metal and paper and new obligations for member states to separately collect bio-waste, hazardous waste produced by households and textile waste – New measures requiring member states to reduce food waste generation at each stage of the food supply chain, monitor and report annually on food waste levels – Requirement for the commission to draw up guidelines to assist and facilitate Member States in the separate collection of hazardous waste fractions produced by households
Improved cooperation with member states for better implementation of EU waste legislation, and combat illicit shipment of end of life vehicles (ELV)	– Support to member states authorities trough networks such as IMPEL – Exchange of good practices on member states campaigns for inspection of ELV treatment facilities, – Reinforced exchange of information regarding certificates of destructions (CoDs) – The on-going review of the End-of-Life Vehicles Directive
Stepping up enforcement of revised Waste Shipment regulation	– An Implementing act adopted 28 July 2016 sets out a preliminary correlation table between customs and waste codes,[q] which will help customs officials to identify more easily potential waste streams – The Commission is fostering the exchange of information with Member States and is exploring the preparation of a set of guidelines to facilitate the interchange of electronic data – An evaluation of the inspection requirements is being carried out and is expected to be finalised in May 2019

(continued)

Table 1.5 (continued)

Actions	What has been delivered
Promotion of industry-led voluntary certification of treatment facilities for key waste/recyclate streams	The promotion of voluntary schemes has been supported with targeted funding from Horizon 2020. The CEWASTE H2020 project[r] aims at understanding existing recovery practices, standards and verification schemes
Initiative on waste to energy in the framework of the Energy Union	The Communication "The role of waste-to-energy in the circular economy"[s] was adopted on 26 January 2017 with the aim to get more energy from less waste
Identification and dissemination of good practices in waste collection systems	– In 2015, a study[t] assessed the separate collection schemes for municipal waste in the capital cities of all EU Member States – The Horizon 2020 project ImpactPapeRec[u] on good practices in paper collection systems was completed in March 2018 – Guidelines on the implementation of separate collection obligations and best practices, in particular focusing on key waste streams, such as plastics, bio-waste and textiles are to be adopted by the end of 2019
Market for secondary raw materials	
Development of quality standards for secondary raw materials (in particular for plastics)	– The Commission has requested the European Committee for Standardisation (CEN) to perform a comprehensive mapping exercise of existing or ongoing standardisation work related to the treatment of waste and the quality of secondary raw materials, in particular for plastics. The report of CEN was delivered in June 2018 – Specific studies are currently focusing on the development of standards for sustainable chemicals and for secondary raw materials
Proposal for a revised fertilisers regulation	The European Parliament and the Council reached on 12 December 2018 a political agreement on a new Regulation on fertilisers
Proposed legislation setting minimum requirements for reused water for irrigation and groundwater recharge	A proposal for a Regulation on minimum requirements for water reuse was adopted on 28 May 2018
Promotion of safe and cost-effective water reuse, including guidance on the integration of water reuse in water planning and management, inclusion of best practices in relevant BREFs, and support to innovation (through the European Innovation Partnership and Horizon 2020) and investments	– In July 2016, the Commission issued Guidelines on Integrating Water Reuse and Water Planning and Management in the context of the Water Framework Directive – Water saving, reuse and recycling is also considered in the development and review of BREFs for relevant (agro)industrial sectors under the scope of the Industrial Emissions Directive – Water reuse was made a top priority area in the European Innovation Partnership (EIP) on Water – Dedicated funding is available in European Regional Development Fund (ERDF), H2020, and LIFE – Support for water reuse infrastructure is made available by the ERDF, the Cohesion Fund and European Agricultural Fund for Rural Development (EARDF)

(continued)

Table 1.5 (continued)

Actions	What has been delivered
Analysis and policy options to address the interface between chemicals, products and waste legislation, including how to reduce the presence and improve the tracking of chemicals of concern in products	– The Commission Communication on options to address the interface between chemicals, product and waste legislations was adopted on 16 January 2018,[v] together with its accompanying staff-working document – A 12-week public consultation closed on 29 October 2018. The Commission Services are now preparing a summary report of the results of the public consultation that should be published in first quarter of 2019
A 12-week public consultation closed on 29 October 2018. The Commission Services are now preparing a summary report of the results of the public consultation that should be published in first quarter of 2019	– In 2020 the Commission will review the Waste Shipment Regulation to assess whether the regulation meets its objectives and is coherent with the general objectives of EU environmental policy, CE and the internal market. Preparatory work for the review includes consultations and workshops with stakeholders. In view to prepare guidelines, a working group of Member States and stakeholders representatives is addressing issues related to electronic data interchange
Further development of the EU raw materials information system	– The Raw Materials Information System (RMIS)[w] launched by JRC in November 2017 – The Raw Materials Scoreboard of indicators[x] – Several Horizon 2020 projects
Sectorial action	
Plastics	
Strategy on plastics in the circular economy	– The EU Strategy for Plastics in a Circular Economy[y] and a Staff Working Document[z] were published on 16 January 2018 in the context of the Circular Economy Package, along with a report on oxo-degradable plastics – The Commission proposed in May 2018 new EU-wide rules to target the 10 single-use plastic items most often found on Europe's beaches and seas – Agreement was also found in December 2018 on the proposed new rules on port reception facilities for the delivery of waste from ships – The Commission also organised a pledging campaign, calling industrial stakeholders to make voluntary pledges to boost the uptake of recycled plastics in products put on the EU market – The Commission has also submitted a file to ECH[aa] in order to seek restrictions for microplastics intentionally added to products
Specific action to reduce marine litter implementing the 2030 Sustainable Development Goals	– On 20 June 2018, the Commission and UN Environment agreed to the 2018 Oceans Roadmap 2 addressing in particular threats of pollution and marine litter – Programmes of measures under the Marine Strategy Framework Directive (MSFD) were submitted in 2016 by MS for reaching good environmental status by 2020 – A series of projects and initiatives under FP7 and H2020 address marine litter, the ecological aspects of microplastics and bio-based solutions – 14 Member States committed resources for the collection of lost fishing gear and marine litter in their respective Operational Programmes for the European Maritime and Fisheries Fund (EMFF)

(continued)

Table 1.5 (continued)

Actions	What has been delivered
Food waste	
Development of a common methodology and indicators to measure food waste	The Commission is elaborating a harmonised methodology to measure food waste at each stage of the food supply chain. The EU Platform on Food Losses and Food Waste contributed to key concepts underlying the methodology for measuring and monitoring food waste. The definition of food waste is now included in the revised waste framework directive, while the methodology is transposed in a delegated act on food waste measurement, to be adopted by end of March 2019. Food waste is also included in the Monitoring Framework of indicators
Stakeholders platform to examine how to achieve SDGs goals on food waste, share best practice and evaluate progress	– The EU Platform on Food Losses and Food Waste was launched in August 2016 – A digital network was set up in 2017 to improve collaboration and exchange amongst Platform members. The Platform has partnered with Horizon 2020 project REFRESH
Clarify relevant EU legislation related to waste, food and feed in order to facilitate food donation and utilisation of former foodstuffs for animal feed	– The Commission published EU Guidelines for the feed use of food no longer intended for human consumption – With support of the EU Platform on Food Losses and Food Waste, the Commission adopted EU guidelines on food donation in October 2017, and the Platform is expected to adopt, early 2019 – An ongoing EU pilot project on food redistribution (2018–2020) will further explore the policy, regulatory and operational frameworks existing in the Member States as well as promote dissemination of the EU food donation guidelines and stakeholder engagement in this regard
Explore options for more effective use and understanding of date marking on food	In February 2018, the European Commission published a market study on date marking practices54 in the EU Technical guidance is currently under preparation, with support of the EU Platform on Food Losses and Food Waste, in order to promote more consistent date marking practices in line with EU date marking rules
Critical raw materials	
Report on critical raw materials and the circular economy	The report on critical raw materials[ab] was published on 16 January 2018
Improve exchange of information between manufacturers and recyclers on electronic products	– The Commission initiated a joint dialogue with the different actors of the value chain (representatives of EEE producers, recyclers, reuse operators) – The "i4R" platform[ac] was launched to allow the exchange of information between producers of electrical and electronic equipment and recyclers of WEEE
European standards for material-efficient recycling of electronic waste, waste batteries and other relevant complex end-of-life products	A first series of standards for the treatment of WEEE has been developed by CENELEC.[ad] The Commission requested the European Standardisation Organisations to further develop European standards for material-efficient recycling of electronic waste and waste batteries with the objective of increasing high-quality recycling of Critical Raw Materials

(continued)

Table 1.5 (continued)

Actions	What has been delivered
Sharing of best practice for the recovery of critical raw materials from mining waste and landfills	– The review of the state of implementation of the Extractive Waste Directive[ae] by Member States was published in 2017 – The Commission's Joint Research Centre is finalizing (due in April 2019) a report gathering best practices on non-critical and critical raw material recovery from mining waste and landfills, as a supporting action for Extractive Waste Management Plans – Two Horizon 2020 projects are completing a secondary raw materials inventory as regards mining waste and landfills
Construction and demolition	
Pre-demolition assessment guidelines for the construction sector	– The outcomes of the study on Pre-demolition & Renovation Waste Audits were released as Guidelines for Assessment of Construction and Demolition Waste Streams prior to the Demolition or Renovation of Buildings and Infrastructures
Voluntary industry-wide recycling protocol for construction and demolition waste	– The EU Construction and Demolition waste management protocol[af] was published in October 2016. Dissemination and communication actions on the protocol have been implemented – In 2017, in parallel to the Pre-demolition assessment guidelines a communication campaign was performed
Core indicators for the assessment of the lifecycle environmental performance of a building, and incentives for their use	– Level(s),[ag] the European reporting framework for sustainable buildings, with its indicators and life cycle tools, has been developed – The test phase, dedicated to test the usefulness and robustness of the different parts of the framework, is now ongoing and will last until summer 2019. A public consultation is foreseen early 2020
Biomass and bio-based materials	
Guidance and dissemination of best practice on the cascading use of biomass and support to innovation in this domain through Horizon 2020	Guidance on cascading use of biomass was published on 16 November 201863
Ensuring coherence and synergies with the circular economy when examining the sustainability of bioenergy under the Energy Union	– The new Renewable Energy Directive contains provisions referring to circular economy and waste hierarchy
Assessment of the contribution of the 2012 Bio-economy Strategy to the circular economy and possible review	– The updated Bioeconomy Strategy and Action plan propose 14 concrete actions – The promotion of bio-based materials and products will be ensured during the development of EU Ecolabel and GPP criteria for new or existing product groups, according to Environmental Footprint results, and in line with available EU standards and technical reports, as well as with the strategic approach for EU Ecolabel and GPP
Innovation and Investments	
Initiative "Industry 2020 and the circular economy" under Horizon 2020	– Two "focus areas" have been dedicated to the circular economy in the Work Programmes 2016–17 and 2018–20 of Horizon 2020, covering call topics specifically developed to address the needs related to a circular economy in a systematic and comprehensive way – The Commission has published an inventory of the projects relevant to the circular economy funded under H2020 between 2016 and 2018[ah]

(continued)

Table 1.5 (continued)

Actions	What has been delivered
Pilot project for "innovation deals" to address possible regulatory obstacles for innovators	– The first call for pilot projects received 32 proposals from 14 different countries. The two selected Innovation Deals focus on: (1) sustainable wastewater treatment and (2) optimising e-vehicle battery usage. Work on the two Innovation Deals is ongoing
Targeted outreach to encourage applications for funding under EFSI, and support the development of projects and investment platforms relevant to the circular economy	– Thematic workshops started in 2016 and continue in 2019 to increase the use of funds for the circular economy through EFSI – To provide advice on funding opportunities the European Investment Advisory Hub (EIAH) was launched, while EUR 100 million are made available via the Circular Bioeconomy Thematic Investment Platform
Targeted outreach and communication activities to assist Member States and regions for the uptake of Cohesion Policy funds for the circular economy	– From 2014 to 2020, cohesion policy allocates around EUR 150 billion to objectives with a direct relevance to the circular economy. The implementation of the national and regional programmes is now fully underway and projects are being selected by the Member States. The Commission offers various mechanisms to help Member States implement the programmes and carry out projects in order to use the available resources in an optimal way. The allocations and expected results are visualised in the new Open Data Platform[ai]
Support to Member States and regions to strengthen innovation for the circular economy through smart specialisation	– About EUR 41 billion are available to implement the so-called smart specialisation strategies of regions and Member States – The Smart Specialisation Platform[aj] and the thematic platforms on energy, agri-food, and industrial modernisation hosted by the JRC help the implementation of those strategies – In 2018, pilot actions have been launched to further support innovation projects
Assessment of the possibility of launching a platform together with the EIB and national banks to support the financing of the	– The Circular Economy Finance Support Platform was launched on 26 January 2017 – A Commission expert group was set-up to coordinate activities regarding the financing of the circular economy and to develop general recommendations
Engagement with stakeholders in the implementation of this action plan through existing fora in key sectors	The European Circular Economy Stakeholder Platform[ak] was launched in March 2017 to foster policy dialogue, to exchange expertise among stakeholders and to identify barriers in relation to the circular economy. A Coordination Group composed of representative from existing networks working on circular economy was set up to multiply the impact of the platform
Support to a range of stakeholders through actions on public-private partnerships, cooperation platforms, support to voluntary business approaches, and exchanges of best practices	– A Smart Specialisation Platform on Industrial Modernisation has been launched in June 2016 to facilitate cross-regional cooperation towards industrial modernisation projects – In 2018 the Pilot Project 'Boosting the circular economy amongst SMEs in Europe' provided online training to SME support organisations and policy advice to regional authorities[al] – The European Resource Efficiency Knowledge Centre was also set up[am] – In 2017, a partnership on Circular Economy within the Urban Agenda for the EU75 was launched – The Commission awarded projects for Urban Innovative Actions on circular economy to 8 cities in October 2017 – The transition towards a circular economy has also been promoted through events organised by the Commission's Representations, through corporate communication campaigns and through Citizens' Dialogues

(continued)

Table 1.5 (continued)

Actions	What has been delivered
Monitoring	
Development of a monitoring framework for the circular economy	A Monitoring Framework of Indicators for the Circular Economy[an] was published on 16 January 2018. The indicators and underlying data are publicly available on a dedicated EUROSTAT website[ao]

Source SWD [35] 90
[a]https://ec.europa.eu/energy/sites/ener/files/documents/com_2016_773.en_.pdf
[b]SWD [35] 91
[c]http://ec.europa.eu/environment/waste/mining/pdf/guidance_extractive_waste.pdf
[d]https://www.ket4sme.eu
[e]http://ec.europa.eu/environment/ecolabel/documents/Report_from_the_Commission.pdf
[f]https://eur-lex.europa.eu/legal-content/EN/TXT/HTML/?uri=CELEX:52017DC0355&from=EN
[g]Contract "Substitution of Chemical Substances of Potential Concern (Phase II)", EASME/COSME/2017/025
[h]https://ec.europa.eu/newsroom/just/item-detail.cfm?item_id=59332
[i]https://ec.europa.eu/info/sites/info/files/communication_11.4.2018.pdf
[j]https://ec.europa.eu/info/sites/info/files/ucp-guidance-en.pdf
[k]http://susproc.jrc.ec.europa.eu/ScoringSystemOnReparability/index.html
[l]http://ec.europa.eu/ecat/
[m]http://ec.europa.eu/environment/eussd/smgp/PEFCR_OEFSR_en.htm
[n]http://ec.europa.eu/environment/gpp/pdf/Buying-Green-Handbook-3rd-Edition.pdf
[o]http://ec.europa.eu/environment/gpp/pdf/cp_european_commission_brochure_en.pdf
[p]https://eur-lex.europa.eu/legal-content/EN/TXT/PDF/?uri=OJ:L:2018:150:FULL
[q]https://publications.europa.eu/en/publication-detail/-/publication/b5a3c820-5546-11e6-89bd-01aa75ed71a1/language-en
[r]https://cordis.europa.eu/project/rcn/218283/factsheet/en
[s]http://ec.europa.eu/environment/waste/waste-to-energy.pdf
[t]http://ec.europa.eu/environment/waste/studies/pdf/Separate%20collection_Final%20Report.pdf
[u]http://impactpaperec.eu/en/home/
[v]https://www.eumonitor.eu/9353000/1/j4nvke1fm2yd1u0_j9vvik7m1c3gyxp/vkl2na5lcqz6/v=s7z/f=/com(2018)32_en.pdf
[w]http://rmis.jrc.ec.europa.eu
[x]https://publications.europa.eu/en/publication-detail/-/publication/117c8d9b-e3d3-11e8-b690-01aa75ed71a1
[y]http://ec.europa.eu/environment/circular-economy/pdf/plastics-strategy.pdf
[z]https://eur-lex.europa.eu/legal-content/EN/TXT/PDF/?uri=CELEX:52018SC0016&from=EN
[aa]https://echa.europa.eu/registry-of-restriction-intentions/-/dislist/details/0b0236e18244cd73
[ab]https://ec.europa.eu/commission/publications/report-critical-raw-materials-and-circular-economy_en
[ac]http://www.i4r-platform.eu
[ad]https://www.cencenelec.eu/news/publications/publications/weee-brochure.pdf
[ae]http://ec.europa.eu/environment/waste/studies/pdf/KH-01-17-904-EN-N.pdf
[af]https://ec.europa.eu/growth/content/eu-construction-and-demolition-waste-protocol-0_en
[ag]https://ec.europa.eu/growth/content/commission-introduces-eu-construction-and-demolition-waste-protocol-1_en
[ah]https://ec.europa.eu/research/environment/index.cfm?pg=output&pubs=thematic
[ai]https://cohesiondata.ec.europa.eu/
[aj]http://s3platform.jrc.ec.europa.eu/
[ak]https://circulareconomy.europa.eu/platform/
[al]http://ec.europa.eu/environment/sme/circular_economy_boost_en.htm
[am]https://www.resourceefficient.eu
[an]http://ec.europa.eu/environment/circular-economy/pdf/monitoring-framework.pdf
[ao]https://ec.europa.eu/eurostat/web/circular-economy/indicators/monitoring-framework

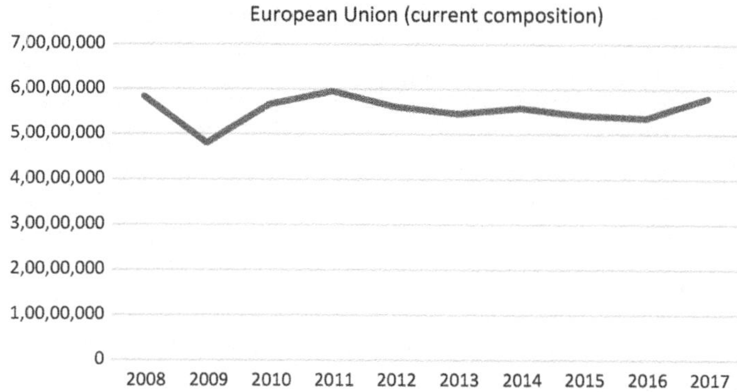

Fig. 1.5 Trade in recyclable raw materials by waste (unite: Tonne) (Euro SDMX Metadata Structure (ESMS) Compiling agency: Eurostat, Statistical Office of the European Union) (http://appsso.eurostat.ec.europa.eu/nui/show.do?dataset=env_wastrd&lang=en)

1.4 The Waste Framework Directive 2008/98/EC

It is nowadays well established the close connection between waste management and the development of policies and strategies supporting the transition to a circular economy, as clearly expressed by the COM (2018) 656 [37]: "*Good waste management is a building block of the circular economy and helps prevent waste from having a negative impact on the environment and health. Proper implementation of the EU's waste legislation will speed up the transition to a circular economy. Local actors have a crucial role in waste management and their involvement in policy development and implementation, as well as support for their activities, is necessary to ensure compliance with EU legislation*".

The influences of the circular economy approach are evident in the last decade EU policies, that have encouraged non-hazardous waste streams, considered valuable resources as potential and important source of raw materials. Overall, cross-border movements of recyclable waste have increased over the last decade[26] and now they present a constant trend (Figs. 1.5 and 1.6).

Waste streams and movements refer to the Waste Framework Directive (WFD) 2008/98/EC, that, a decade after its enactment, is still the undisputed legislative framework for the handling of waste in the Community.[27] The Directive, in fact, dealing with contents and methods, surely constitutes an essential reference for national authorities and for different types of stakeholders dealing with any action concerning waste [39]. The Waste Framework Directive 2008/98/EC, recently amended by the

[26]Trade in recyclable raw materials by waste (env_wastrd) Reference Metadata in Euro SDMX Metadata Structure (ESMS) Compiling agency: Eurostat, Statistical Office of the European Union.

[27]See: European Commission Directorate-General Environments, *Guidance on the interpretation of key provisions of Directive 2008/98/EC on waste* [38].

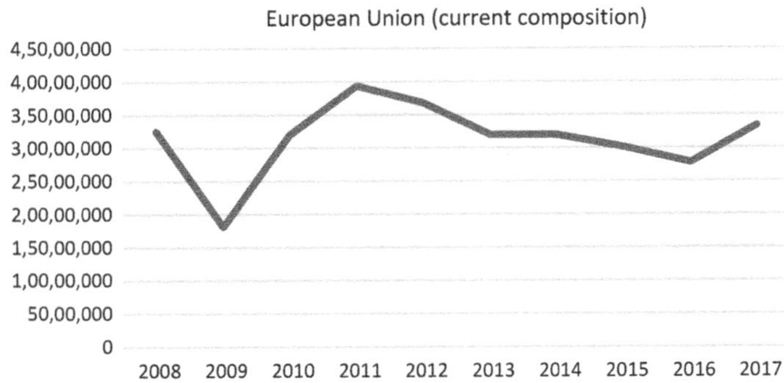

Fig. 1.6 Trade in recyclable raw materials by waste (unite: Thousand euro)

DIRECTIVE 2018/851,[28] is the fundamental legislative document on waste at the EU level, transposed into the national legislation of the Member States by means of separate legal acts. With this Directive, the European Union, aiming to remove the link between economic growth and the production of waste, has modified the legal framework related to waste. Member States are required to prepare waste management plans taking into account the type, the quantities, the sources and the systems of waste collection. The Waste Framework Directive 2008/98, besides, providing common criteria and goals to all the Member States for a more and more effective management of waste, has also placed the basis for the launch and the development of circular economy, orienting actions both at the political and at the operational levels.

In short, the waste Framework Directive 2008/98/EC establishes some basic waste management principles and a legal framework for treating waste in the EU with the aim to protect the environment and human health, to prevent or reduce the generation of waste, to promote recovery and recycling techniques, to reduce pressure on resources, to boost the transition to a circular economy and to guarantee the Union's long-term competitiveness. At this aim, the Directive, among others:

- specifies the basic concepts and definitions related to waste management, such as definitions of waste, re-use, recycling, recovery, prevention, backfilling (D. 2008/98 art. 3 emended by d. 2018/851);
- describes the waste hierarchy (prevention, preparing for re-use, recycling, other recovery, e.g. energy recovery and disposal) (D. 2008/98 art. 4), that represents a common reference for the measures the Member States shall take and, for each of them, establishes goals and targets (art.9-12);
- distinguishes between waste and by-products (D. 2008/98 art. 5 emended by D. 2018/851), specifying the criteria that allow to deal with substances or objects, not considered as waste, even if resulting from a production process whose primary aim is not the production of those items. By-product is a fundamental concept for

[28]Directive (EU) 2018/851 [36] of the European Parliament and of the council of 30 May 2018 amending Directive 2008/98/EC on waste.

circular economy, that allows to recognize the economic and environmental value of items, that can be placed on the market, avoiding to increase the waste streams;
- explains the end-of-waste criteria, that is the conditions under which a waste ceases to be waste (D. 2008/98 art. 6 emended by D. 2018/851) when has undergone a recycling or other recovery operation[29];
- as far as the list of waste (D. 2008/98 art. 7 emended by d. 2018/851), refers to the Decision 2000/532/EC [40][30] and defines criteria for the updating of the list;
- introduces the concept of 'extended producer responsibility', according to which the manufacturers—any natural or legal person who professionally develops, manufactures, processes, treats, sells or imports products (producer of the product)—are requested to accept and dispose of products returned after use. These measures may include the obligation to provide publicly available information as to the extent to which the product is re-usable and recyclable;
- establishes the 'polluter pays principle' whereby, according to legislative or non-legislative measures, the original waste producer must take charge of waste management by performing various activities (D. 2008/98 art. 8), that is the acceptance of returned products and of the waste that remains after those products have been used, as well as the subsequent management of the waste and financial responsibility for such activities. Directive 2018/851 adds an articulated list of requirements for extended producer responsibility schemes, that should be considered by the Member States;
- establishes the responsibility for waste management (D. 2008/98 art. 15), specifying that Member States shall take the necessary measures to ensure that any original waste producer, or other holder, carries out the treatment of waste himself or has the treatment handled by a dealer or an establishment or undertaking which carries out waste treatment operations or arranged by a private or public waste collector;
- requires that Member States adopt waste management plans and waste prevention programmes.

The Waste Framework Directive contained innovative elements for the development of circular economy scenarios and has constituted an important conceptual basis

---

[29]Besides, the Directive 2018/851 puts a further condition: "*The natural or legal person who: (a) uses, for the first time, a material that has ceased to be waste and that has not been placed on the market; or (b) places a material on the market for the first time after it has ceased to be waste, shall ensure that the material meets relevant requirements under the applicable chemical and product related legislation*". The conditions laid down in paragraph 1 have to be met before the legislation on chemicals and products applies to the material that has ceased to be waste.
The Directive stresses the importance of monitoring the development of national end-of-waste criteria: "*The Commission shall monitor the development of national end-of-waste criteria in Member States, and assess the need to develop Union-wide criteria on this basis. To that end, and where appropriate, the Commission shall adopt implementing acts in order to establish detailed criteria on the uniform application of the conditions laid down in paragraph 1 to certain types of waste*".

[30]Commission Decision 2000/532/CE [41] of 3 May 2000 replacing Decision 94/3/EC establishing a list of wastes pursuant to Article 1(a) of Council Directive 75/442/EEC on waste and Council Decision 94/904/EC establishing a list of hazardous waste pursuant to Article 1(4) of Council Directive 91/689/EEC on hazardous waste.

for creating, at the Member State level, the conditions for the progressive integration between waste management and industrial policies.

These aims have been further affirmed and specified in the subsequent COM (2014) 397 [42], that stems from intention of the Community to boost with more force the evolution of the policies of waste management towards the circular economy approaches, jointed with the objectives of the Resource Efficiency Roadmap[31] and 7th Environment Action Programme,[32] including full implementation of the waste hierarchy in all Member States. COM (2014) 397 [42] proposes to emendate the Directive also following the observation of a generalized delay in the implementation as well as a gap amongst[33] the Member States and large divergences in terms of waste management performances.

Three main studies,[34] carried out to monitor the impact of the waste legislation, delivered an impact assessment highlighting the technological and socio-economic benefits related to the implementation and further development of EU waste legislation.

The successor COM (2015) 595 [44] confirmed the contents of the COM (2014) 397 [42] and the impact assessment, reaching the conclusion that a combination of options will bring some main benefits:

- "*Administrative burden reduction in particular for small establishments or undertakings, simplification and better implementation including by keeping targets 'fit for purpose';*
- *Job creation—more than 170.000 direct jobs could be created by 2035, most of them impossible to delocalize outside the EU;*
- *GHG emission reduction—more than 600 millions of tons of green house gas could be avoided between 2015 and 2035;*
- *Positive effects on the competitiveness of the EU waste management and recycling sectors as well as on the EU manufacturing sector (better extended producer responsibility schemes, reduced risks associated with raw material access);*
- *Reinjection into the EU economy of secondary raw materials which in turn will reduce the dependency of the EU on raw materials imports.*" [44].

[31]COM (2011) 571 [43].

[32]Decision No 1386/2013/EU of the European Parliament and of the Council of 20 November 2013 on a General Union Environment Action Programme to 2020 'Living well, within the limits of our planet'.

[33]"*In 2011, while six Member States landfilled less than 3% of their municipal waste, 18 lost resources by landfilling over 50%, with some exceeding 90% of landfilling. This shows large divergences in terms of waste management performances which need to be redressed as a matter of urgency*" COM (2014) 397 [42].

[34]An Impact Assessment Steering Group was created on 16 April 2012 for the preparation of the Impact Assessmen. Activating an external consultation, a list of issues to be tackled was developed by the Commission and the first interviews with key stakeholders started in February 2013, after an online public consultation was lauched. 670 responses were submitted, reflecting high public concern about the waste management situation in the EU and highlighting high expectations for EU action. An impact assessment report and an executive summary were published. The impact assessment evaluated the main environmental, social and economic impacts of various policy options to improve the waste management records in the EU.

Starting from these assumptions and aiming "*to set clear long-term policy objectives in order to guide measures and investments, notably by preventing the creation of structural overcapacities for the treatment of residual waste and lock-ins of recyclable materials at the lower levels of the waste hierarchy*",[35] Directive 2018/851 has been issued. The recent Directive makes some amendments in order to update the Directive 2008/98/EC on the basis of the advancements of knowledge and the awareness deriving from the many initiatives, developed in the last decade, in the field of circular economy. These amendments provide, among other things, to:

- insert a set of new definitions (non-hazardous waste, municipal waste; construction and demolition waste; food waste; material recovery; backfilling; extended producer responsibility scheme);
- increase targets for preparing for re-use and recycling of waste[36];
- remove substances intended for animal feed from the scope of Directive 2008/98/EC;
- define criteria and requirements for which certain waste can cease to be waste;
- set out exemptions for separation of waste collection;
- establish bio-waste separation;
- establish household hazardous waste collection;
- update record keeping requirements;
- in order to avoid waste treatments which lock in resources at the lower levels of the waste hierarchy and to reinforce the forms of reprocessing of waste in the perspective of the circular economy, introduces the term of secondary raw materials[37] and underlines the conditions necessary for supporting their market [38];

[35]Directive (EU) 2018/851 [36].

[36]Dir. 2008/98 establishes targets related to preparing for re-use and the recycling: "*By 2020, the preparing for re-use and the recycling of waste materials such as at least paper, metal, plastic and glass from households and possibly from other origins as far as these waste streams are similar to waste from households, shall be increased to a minimum of overall 50% by weight; by 2020, the preparing for re-use, recycling and other material recovery, including backfilling operations using waste to substitute other materials, of non-hazardous construction and demolition waste excluding naturally occurring material defined in category 17 05 04 in the list of waste shall be increased to a minimum of 70% by weight*".

To these targets the Directive 2018/851 adds three new forward-looking targets, related to municipal waste: "*by 2025, the preparing for re-use and the recycling of municipal waste shall be increased to a minimum of 55% by weight; by 2030, the preparing for re-use and the recycling of municipal waste shall be increased to a minimum of 60% by weight; by 2035, the preparing for re-use and the recycling of municipal waste shall be increased to a minimum of 65% by weight*".

[37]"*A definition of material recovery should be introduced to cover forms of recovery other than energy recovery and other than the reprocessing of waste into materials used as fuels or other means to generate energy. It includes preparing for re-use, recycling and backfilling and other forms of material recovery such as the reprocessing of waste into secondary raw materials for engineering purposes in construction of roads or other infrastructure. Depending on the specific factual circumstances, such reprocessing can fulfil the definition of recycling if the use of materials is based on proper quality control and meets all relevant standards, norms, specifications and environmental and health protection requirements for the specific use*" (Directive 2018/851).

[38]*In order to provide operators in markets for secondary raw materials with more certainty as to the waste or non- waste status of substances or objects and to promote a level playing field, it is*

- introduces the Early Warning Reports in order to list Member States at risk of not attaining the targets within the respective deadlines and to deliver appropriate recommendations for the Member States concerned.

Considering this last aspect, it is significant to underline the Commission's intention to strengthen the monitoring actions. Early Warning Reports will be developed in order to provide an estimation of whether the targets are likely to be achieved by the stipulated deadline and to suggest appropriate Priority Actions to the countries at greatest risk of missing the target.

In this regard, both for understanding the characteristics of this kind of reports and for having an overview of the Member States behaviors, related to the targets of the Waste Framework Directive, it is interesting to mention the Early Warning Report developed by Eunomia Research & Consulting [45], commissioned by the European Commission. The Report provides an overview of the approach taken for the study (Fig. 1.7), identifies (according to the defined assessment methods) the Member States with a high risk of failing to meet the 50% target in 2020[39] (Table 1.6), summarizes the key issues found and finally suggests Priority Actions.

important that Member States take appropriate measures to ensure that waste that has undergone a recovery operation is considered to have ceased to be waste if it complies with all the conditions (Directive 2018/851).

[39]The study assumes the target of 50% preparation for reuse and recycling target set out in Article 11 of the Waste Framework Directive (Directive 2008/98/EC) even if the targets proposed by the COM [44] 595, and established by the recent Directive 2018/851, are higher.

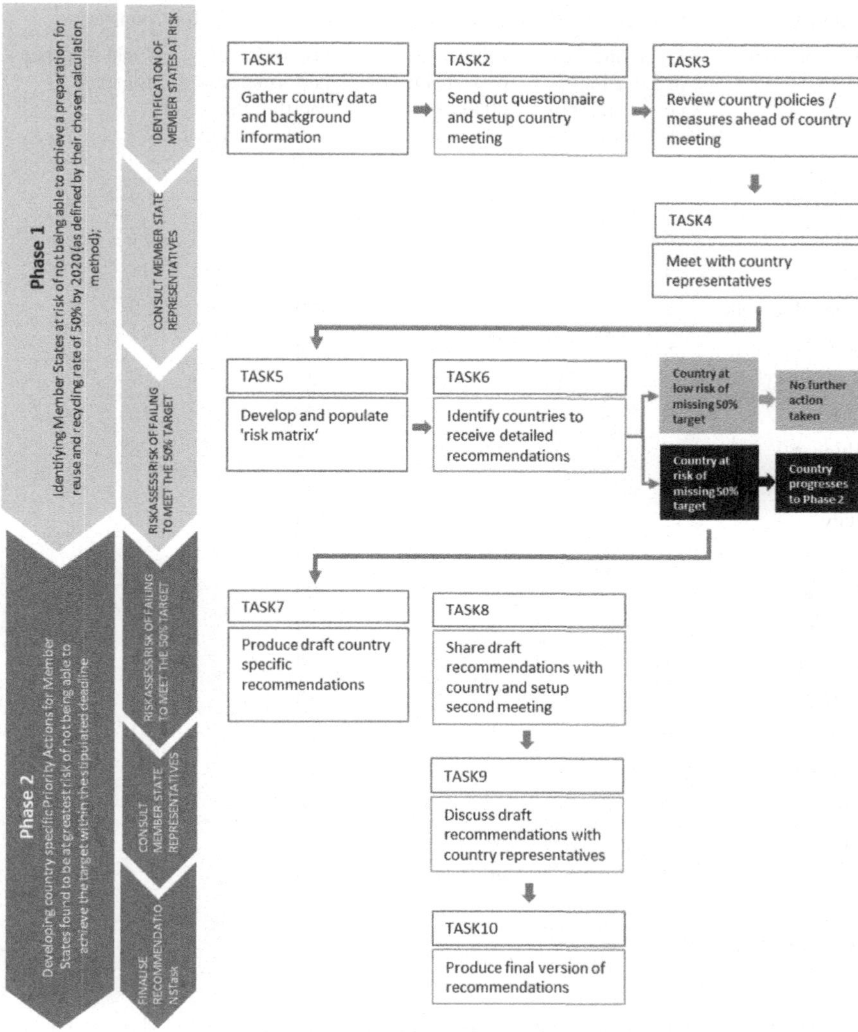

Fig. 1.7 Two-phased approach for the Identification of Member States at Risk of Non-Compliance with the 2020 Target of the Waste Framework Directive. *Source* Eunomia [45]

The study highlights a critical aspect, that should be carefully considered in relation to the information provided by Member States and the accuracy of European waste statistics. By comparing Eurostat data with the estimate from the Member States Waste data, significant issues with many of the datasets still remain, owing

Table 1.6 Detailed Assessment of Likely Recycling Performance in 2020 for Countries

Member State	Likely increase in recycling performance by 2020 (%)	Likely recycling performance by 2020 (%)	Proportion of performance shortfall likely to be closed by 2020 (%)	Final Distance to target
Bulgaria	12	41	58	9
Croatia	8	33	32	17
Cyprus	4	32	18	18
Czech Republic	2	51	100	0
Estonia	6	39	34	11
Finland	3	43	26	7
Greece	9	35	36	15
Hungary	0	42	0	8
Ireland	7	52	100	0
Italy	12	58	100	0
Latvia	7	34	30	16
Lithuania	5	50	100	0
Malta	12	28	35	22
Poland	21	47	88	3
Portugal	5	37	28	13
Romania	6	19	16	31
Slovakia	7	27	23	23
Slovenia	0	37	0	13
Spain	6	37	31	13
United Kingdom	2	46	35	4

Source Eunomia [45]

to at least two main causes, that is: insufficient verification of the data at EU and national level and a lack of incentives for accurate data reporting [46].

The study allows to obtain a general overview of the common issues and barriers, at various levels (Table 1.7), to achieving targets [47–49].

Table 1.7 Common issues and risk of Non-Compliance with the 2020 Target of the Waste Framework Directive

Issues	Description
Common issues—national level	
National targets not cascaded to regional/municipal level or minimal consequence of failure	Many Member States have national recycling targets in place in order to be compliant with the transposition of the Waste Framework Directive. A range of other policies are also often in place, such as waste taxes or bans. However, the presence of national targets does not necessarily mean that action will take place on the ground. (…) Where targets are maintained only at the national level, there is likely to be little effect without other strong policies in place. National targets could be cascaded down to the municipal level but are often not
Low costs of disposal	Many Member States do have landfill bans or taxes, but many still do not, or where taxes are in place the rates are still too low to provide a clear economic justification for investment in alternative reuse, recycling and composting systems
National policies requiring food waste separation too loosely worded and will not result in significant changes	Many policies aiming to place obligations on waste producers to sort food waste are too loosely worded, and would be unlikely to see significant changes
Lack of enforcement of policies	Several Member States have good policies in place, but are not being fully effective as they are not being enforced. This is in the main due to a lack of political will to prioritise changes to local government services
Regulatory uncertainty/continuous small changes	Regulatory uncertainty (absence of clear regulation, or regulation is changing frequently due to changes political conditions) provides a major barrier to investment in the sector. Besides, national and local administrations often lack the resourcing and knowledge to produce robust administrative and regulatory frameworks for the country
Inadequate appraisal of best practice options in policy design	In many Member States national policy often seems to be implemented without a thorough understanding of the full range of best practice options that might be suitable in the country

(continued)

Table 1.7 (continued)

Issues	Description
No/insufficient frameworks for consistency, leading to highly variable, sometimes poorly implemented systems	Many countries need to balance central versus local government control of services, and regularly many of the decisions about implementation are left solely to the municipalities
Outdated data capture systems providing uncertainty about existing performance	Only some Member States have invested in modern digital waste information systems, so it is difficult to report accurate data related to the calculation of the household and similar waste recycling rate under the Waste Framework Directive. Consequently, the uncertainty about the existing level of performance deters investments necessary to close the possible gap
Data on products placed on the market through Extended producer responsibility schemes being underreported	Cross-checks of different data source highlights inaccuracies in the source data, especially for packaging recycling rates
Municipal level	
No integration between Extended producer responsibility schemes and Member State waste services/fragmentation of responsibilities	The lack of integration causes inefficient service design, as services cannot be optimised across dry recycling, food and garden and mixed waste collections. Moreover, where recycling schemes are completely separate from those responsible for collection and treatment of mixed wastes, the schemes don't see financial benefits from reduced mixed waste arising
Minimal/no capacity to design and deliver selective collection systems	Local governments are often unprepared for service changes that should take into account a range of factors such as consumer behaviour, the value of secondary materials market, effective communications campaigns, robust procurement and contract management
High proportion of housing stock as apartments/challenges in collection of recycling	Developing systems which deliver high levels of capture of recycling from the housing stock as apartments will be a critical part of increasing recycling rates in many Member States
Packaging recycling rates in some cases appear overestimated, reducing the financial contributions to the system and effort from producers	The packaging element of service provision can be underfunded

(continued)

Table 1.7 (continued)

Issues	Description
Overreliance on EU Funds	For several Member States there has been an overreliance on using EU Funds for development of infrastructure, leaving no longer term funding plan in place. This is most relevant for collection and recycling infrastructure, which tends to be lower in capital cost and higher in operational costs
Lack of effective communication campaigns	If citizens are requested to pay higher fees for services, it is important to properly communicate the benefits for the country (and planet)
Waste industry can lack required skills and competences	The waste industry may not have the experience to design and operate the required services to a high standard, especially in those countries which don't have international waste companies in the market that can bring in expertise from other countries
Citizens	
No financial or other incentive to separate recyclables (lack of participation)	Often there are no measures that provide specific incentives to citizens
Inconvenient selective collection systems (i.e. not door-to-door or near entry)	Recycling containers at the end of the local street is a low cost way of meeting low recycling targets
Lack of public understanding of systems and need to minimise contamination	In many cases the public are unaware of the issues with contamination of recycling streams particularly, contamination of food waste from plastics or segregated plastics stream with incorrect polymers or other contaminants. Moreover, many citizens are sceptical of the final destination of segregated recycling streams, particularly when exported

Source Eunomia [45]

References

1. Jesus Pacheco DA, De Caten CS, Jung CF, Navas HV, Cruz-Machado VA et al (2018) Eco-innovation determinants in manufacturing SMEs from emerging markets: systematic literature review and challenges. J Eng Tech Manag 48:44–63
2. De Jesus A, Mendonça S (2018) Lost in transition? Drivers and barriers in the eco-innovation road to the circular economy. Ecol Econ 145:75–89
3. Ghisetti C, Montresor S (2018) Design and eco-innovation: micro-evidence from the Eurobarometer survey. Ind Innov 1–34
4. Lee CH, Wu KJ, Tseng ML (2018) Resource management practice through eco-innovation toward sustainable development using qualitative information and quantitative data. J Clean Prod 202:120–129
5. Sullivan K, Thomas S, Rosano M (2018) Using industrial ecology and strategic management concepts to pursue the sustainable development goals. J Clean Prod 174:237–246

6. Zaoual AR, Lecocq X (2018) Orchestrating circularity within industrial ecosystems: lessons from iconic cases in three different countries. Calif Manag Rev 60(3):133–156
7. Meyer B et al (2011) Macroeconomic modelling of sustainable development and the links between the economy and the environment. Study for the European Commission (DG Environment)
8. Tseng ML, Tan RR, Chiu AS, Chien CF, Kuo TC (2018) Circular economy meets industry 4.0: can big data drive industrial symbiosis? Resour Conserv Recycl 131:146–147
9. COM (2005) 666 Communication from the commission to the council, the European Parliament, the European economic and social committee and the committee of the regions taking sustainable use of resources forward: a thematic strategy on the prevention and recycling of waste
10. COM (2011) 13, Report from the commission to the European Parliament, the council, the European economic and social committee and the committee of the regions on the thematic strategy on the prevention and recycling of waste
11. COM (2003) 301 Communication from the commission: towards a thematic strategy on the prevention and recycling of waste
12. EU Waste Policy (2011) The story behind the strategy, EC Report
13. Decision No 1600/2002/EC of the European Parliament and of the council of 22 July 2002 laying down the sixth community environment action programme
14. Deters H (2019) European environmental policy at 50: Five decades of escaping decision traps? Environmental Policy and Governance
15. Endl A, Berger G (2014) The 7th environment action programme: reflections on sustainable development and environmental policy integration, ESDN Quarterly Report, No. 32
16. Krämer L (2019) Global environmental challenges and the EU. ERA Forum, Springer, Berlin, Heidelberg
17. Machin A (2019) Changing the story? The discourse of ecological modernisation in the European Union. Environ Polit 28(2):208–227
18. European Commission (2010) Being wise with waste: the EU's approach to waste management
19. COM (2011) 531, Communication from the commission to the European Parliament, the council, the European economic and social committee and the committee of the regions. The sixth community environment action programme, final assessment
20. Decision No 1386/2013/EU of the European Parliament and of the council of 20 November 2013 on a general union environment action programme to 2020 'Living well, within the limits of our planet'
21. COM (2014) 398, Towards a circular economy: a zero waste pro-gramme for Europe
22. EEA (2016) Circular economy in Europe. Developing the knowledge base, Report
23. EEA (2012) Movements of waste across the EU's internal and external borders, Report
24. EEA (2012) European innovation partnership on raw materials, raw materials scoreboard, European Union, 2016
25. Ellen MacArthur Foundation, the McKinsey centre for business and environment and the Stiftungsfonds für Um-weltökonomie und Nachhaltigkeit (SUN), growth within: a circular economy vision for a competitive Europe, report 2015
26. COM (2014) 398, Communication from the commission to the European Parliament, the council, the European economic and social committee and the committee of the regions. Towards a circular economy: a zero waste programme for Europe
27. COM (2014) 440, Communication from the commission to the European Parliament, the council, the European economic and social committee and the committee of the regions. Green action plan for SMEs. Enabling SMEs to turn environmental challenges into business opportunities
28. COM (2018) 29 communication from the commission to the European Parliament, the council, the European economic and social committee and the committee of the regions on a monitoring framework for the circular economy
29. COM (2015) 614, Communication from the commission to the European Parliament, the council, the European economic and social committee and the committee of the regions. Closing the loop—an EU action plan for the Circular Economy

30. COM (2015) 614 ANNEX 1 to the Communication from the commission to the European Parliament, the council, the European economic and social committee and the committee of the regions closing the loop—an EU action plan for the circular economy
31. Directive 2009/125/EC of the European Parliament and of the council of 21 October 2009 establishing a framework for the setting of ecodesign requirements for energy-related products
32. European Parliament Resolution of 9 July 2015 on resource efficiency: moving towards a circular economy (2014/2208(INI)
33. Eunomia Research & Consulting (2017) ENT Environment and Management, Ekokonsultacijos, final report for DG environment under framework contract ENV.C.2/FRA/2013/0023, study on waste statistics—a comprehensive review of gaps and weaknesses and key priority areas for improvement in the EU waste statistics
34. COM (2019) 190, Report from the commission to the European Parliament, the council, the European economic and social committee and the committee of the regions on the implementation of the circular economy action plan
35. SWD (2019) 90 Commission staff working document accompanying the document report from the commission to the european parliament, the council, the european economic and social committee and the committee of the regions on the implementation of the circular economy action plan
36. Directive (EU) 2018/851 of the European Parliament and of the council of 30 May 2018 amend-ing Directive 2008/98/EC on waste
37. COM (2018) 656 Report from the commission to the European Parliament, the council, the European economic and social committee and the committee of the regions on the implementation of EU waste legislation, including the early warning report for member states at risk of missing the 2020 preparation for re-use/recycling target on municipal waste
38. European Commission Directorate-General Environments (2012) Guidance on the interpretation of key provisions of Directive 2008/98/EC on waste
39. Pires A, Martinho G, Rodrigues S, Gomes MI (2019) Preparation for reusing, recycling, recovering, and land-filling: waste hierarchy steps after waste collection. In: Sustainable Solid Waste Collection and Management, Springer, Cham
40. Decision 2000/532/EC
41. Commission Decision 2000/532/CE of 3 May 2000 replacing decision 94/3/EC establishing a list of wastes pursuant to Article 1(a) of council directive 75/442/EEC on waste and council decision 94/904/EC establishing a list of hazardous waste pursuant to Article 1(4) of council directive 91/689/EEC on hazardous waste
42. COM (2014) 397 Proposal for a directive of the European Parliament and of the council amending directives 2008/98/EC on waste, 94/62/EC on packaging and packaging waste, 1999/31/EC on the landfill of waste, 2000/53/EC on end-of-life vehicles, 2006/66/EC on batteries and accumulators and waste batteries and accumulators, and 2012/19/EU on waste electrical and electronic equipment
43. COM (2011) 571 Decision No 1386/2013/EU of the European Parliament and of the Council of 20 November 2013 on a General Union Environment Action Programme to 2020 'Living well, within the limits of our planet'
44. COM (2015) 595 Proposal for a directive of the European Parliament and of the council amending directive 2008/98/EC on waste
45. Eunomia (2018) Study to identify member states at risk of non-compliance with the 2020 target of the waste framework directive and to follow-up phase 1 and 2 of the compliance promotion exercise. Report for the European Commission, DG environment, waste management and secondary materials unit
46. Eunomia (2017) Recycling—who really leads the world?
47. Bahn-Walkowiak B, Steger S (2015) Resource targets in Europe and world-wide: an overview. Resources 4:597–620
48. Martin M, Harris S (2018) Prospecting the sustainability implications of an emerging industrial symbiosis network. Resour Conserv Recycl 138:246–256
49. United Nations (2015) Transforming our world: the 2030 agenda for sustainable development
50. European Innovation Partnership (EIP) on raw materials, raw materials scoreboard (2016)

Websites

51. http://ec.europa.eu/ecat/
52. http://ec.europa.eu/environment/circular-economy/pdf/monitoring-framework.pdf
53. http://ec.europa.eu/environment/circular-economy/pdf/plastics-strategy.pdf
54. http://ec.europa.eu/environment/enveco/studies_modelling/pdf/report_macroeconomic.pdf 4
55. http://ec.europa.eu/environment/eussd/smgp/PEFCR_OEFSR_en.htm
56. http://ec.europa.eu/environment/gpp/pdf/Buying-Green-Handbook-3rd-Edition.pdf
57. http://ec.europa.eu/environment/gpp/pdf/cp_european_commission_brochure_en.pdf
58. http://ec.europa.eu/environment/newprg/index.htm
59. http://ec.europa.eu/environment/resource_efficiency/targets_indicators/scoreboard/index_en.htm
60. http://ec.europa.eu/environment/sme/circular_economy_boost_en.htm
61. http://ec.europa.eu/environment/waste/studies/pdf/KH-01-17-904-EN-N.pdf
62. http://ec.europa.eu/environment/waste/studies/pdf/Separate%20collection_Final%20Report.pdf
63. http://ec.europa.eu/environment/waste/waste-to-energy.pdf
64. http://impactpaperec.eu/en/home/
65. http://rmis.jrc.ec.europa.eu
66. http://s3platform.jrc.ec.europa.eu/
67. http://susproc.jrc.ec.europa.eu/ScoringSystemOnReparability/index.html
68. http://www.i4r-platform.eu
69. https://circulareconomy.europa.eu/platform/
70. https://cohesiondata.ec.europa.eu/
71. https://cordis.europa.eu/project/rcn/218283/factsheet/en
72. https://ec.europa.eu/commission/publications/report-critical-raw-materials-and-circular-economy_en
73. https://ec.europa.eu/energy/sites/ener/files/documents/com_2016_773.en_.pdf
74. https://ec.europa.eu/eurostat/tgm/refreshTableAction.do?tab=table&plugin=1&pcode=cei_cie010&language=en
75. https://ec.europa.eu/eurostat/tgm/table.do?tab=table&init=1&language=en&pcode=cei_srm030&plugin=1
76. https://ec.europa.eu/eurostat/web/circular-economy
77. https://ec.europa.eu/growth/content/commission-introduces-eu-construction-and-demolition-waste-protocol-1_en
78. https://ec.europa.eu/growth/content/eu-construction-and-demolition-waste-protocol-0_en
79. https://ec.europa.eu/info/sites/info/files/communication_11.4.2018.pdf
80. https://ec.europa.eu/info/sites/info/files/ucp-guidance-en.pdf
81. https://ec.europa.eu/newsroom/just/item-detail.cfm?item_id=59332
82. https://ec.europa.eu/research/environment/index.cfm?pg=output&pubs=thematic
83. https://echa.europa.eu/registry-of-restriction-intentions/-/dislist/details/0b0236e18244cd73
84. https://eur-lex.europa.eu/legal-content/EN/TXT/PDF/?uri=CELEX:52018SC0016&from=EN
85. https://eur-lex.europa.eu/legal-content/EN/TXT/PDF/?uri=OJ:L:2018:150:FULL
86. https://publications.europa.eu/en/publication-detail/-/publication/1ee65e21-9ac4-11e6-868c-01aa75ed71a1
87. https://publications.europa.eu/en/publication-detail/-/publication
88. https://www.cencenelec.eu/news/publications/publications/weee-brochure.pdf
89. https://www.eumonitor.eu/9353000/1/j4nvke1fm2yd1u0_j9vvik7m1c3gyxp/vkl2na5lcqz6/v=s7z/f=/com(2018)32_en.pdf
90. https://www.resourceefficient.eu
91. https://www.eea.europa.eu/data-and-maps/figures/trade-in-selected-waste-materials

Chapter 2
Construction and Demolition Waste

Abstract The chapter deals with the subject of Construction and Demolition Waste (CDW) and the role of recycling and reuse in reducing costs and negative environmental impacts, related to the extraction, processing and production of construction materials. The chapter analyses the characteristics of the CDW and investigates the key actions for reducing the general environmental impacts of the construction sector related to waste production, focusing on barriers and drivers for the development of a market of secondary materials and building components.

Keywords Construction and demolition waste · Secondary materials · Construction sector

2.1 The Impacts of the Construction Sector

The construction sector uses resources and produces waste (Fig. 2.1) more than any other industrial sector: "*By improving resource efficiency in constructing and use of infrastructure and buildings, the EU can influence 42% of its final energy consumption, about 35% of its greenhouse gas emissions and more than 50% of all extracted materials, and save up to 30% water. Economically, construction is one of Europe's largest industrial sectors, with an annual turnover exceeding 1200 billion Euros, and activities that account for 10.4% of the EU GDP. 7.2% of the EU workforce is directly in the building and construction sector. The aggregated impacts of housing and infrastructure account for around 15–30% of all environmental pressures of European consumption. Housing and infrastructure contributes approximately 2.5 tonnes of* CO_2 *equivalent of greenhouse gasses per capita per year. 40% of these GHG emissions are directly associated with heating and hot water for private households. The construction of buildings and other infrastructures contributes another 30% of the total emissions*" SEC (2011) 1067 [1].

This chapter is authored by Cinzia Talamo, Politecnico di Milano, Department of Architecture, Built Environment and Construction Engineering, Milan, Italy. e-mail: cinzia.talamo@polimi.it.

M. Migliore et al., *Strategies for Circular Economy and Cross-sectoral Exchanges for Sustainable Building Products*, Springer Tracts in Civil Engineering,
https://doi.org/10.1007/978-3-030-30318-1_2

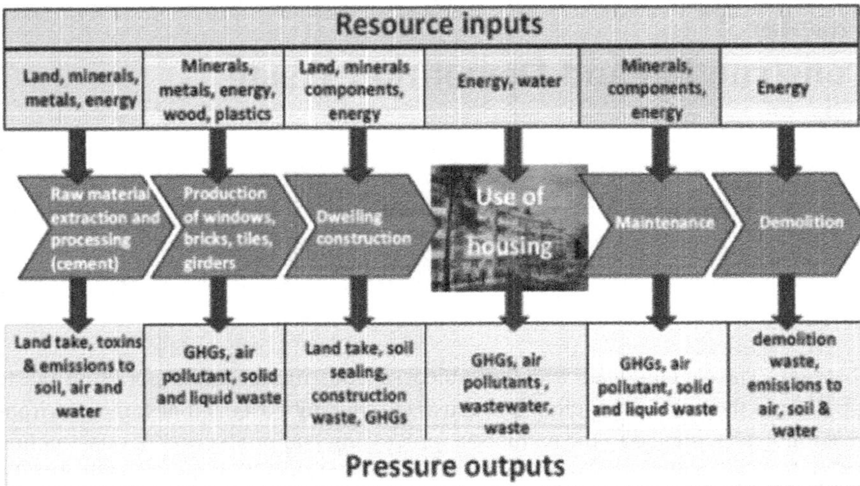

Fig. 2.1 The resources used along the value chain of construction. *Source* SEC (2011) 1067 [1]

The European Construction Sector Observatory [2] estimates that: "*One of the most resource- and waste-intensive economic activities is construction. The construction sector produced 923 million tonnes of waste in 2016, which in terms of volume is the largest waste stream in the EU, representing 30% of all waste generated. Construction and demolition waste (CDW) refers to the waste generated from general construction activities and includes concrete, bricks, gypsum, wood, glass, metal, plastic, solvents, asbestos and excavated soil. The recycling and reuse of CDW components has high potential in reducing construction costs and negative environmental impacts, related to the extraction, processing and production of construction materials*". Many studies [3–9] and statistics (Eurostat) compare present and forecast future input and output of construction materials flows. Whatever the observed level (cities and neighborhoods, buildings, assemblies and components), all they highlight the impacts of the construction sector. It puts the most pressure on the natural environment for both use of raw materials (Table 2.1) and generation of waste (Fig. 2.2) and consequently it has a pivotal role in transitioning to circular economy [10].

Focusing on the strategies aiming to reduce the general environmental impacts of the construction sector, surely reuse, recycling and remanufacturing are a triad of actions strongly driven by the Waste Directive (2008/98/EC), that establishes the well-known waste hierarchy (see Chap. 1) (prevention; preparing for re-use; recycling; other recovery, e.g. energy recovery; disposal).

In-fact, construction and demolition waste (CDW)[1] is the most substantial waste stream in the EU by weight, accounting for over 800 million tonnes per year, i.e.

[1]"*Construction and demolition waste refers to waste that results from construction and demolition activities in a general way, it also includes waste arising from minor do-it-yourself construction and demolition activities within private households. Construction and demolition waste should*

Table 2.1 Material flow accounts in raw material equivalents by final uses of products—modelling estimates by Eurostat

CPA08/TIME	2009	2010	2011	2012	2013	2014	2015	2016
Total CPA products	7.691.982,095	7.523.937,908	7.737.632,12	7.197.717,101	7.078.241,224	7.224.583,121	7.224.856,192	7.244.228,52
Products of agriculture, hunting and related services	423.952,453	397.766,575	426.900,188	421.913,758	404.651,769	455.098,218	404.926,957	414.516,315
Products of forestry, logging and related services	78.932,193	87.077,282	105.254,9	97.958,072	103.792,984	126.626,654	116.223,164	133.210,356
Fish and other fishing products; aquaculture products; support services to fishing	6.737,749	6.759,687	6.511,886	6.465,293	6.885,106	6.478,896	7.322,645	7.002,6
Mining and quarrying	177.358,22	140.512,815	150.269,505	148.524,455	153.665,609	149.067,449	136.033,367	165.060,491
Food, beverages and tobacco products	891.426,7	859.779,151	865.541,157	853.022,657	898.065,741	941.137,464	917.008,807	940.174,887
Textiles, wearing apparel, leather and related products	84.222,723	93.487,988	97.267,698	85.907,682	86.095,756	93.481,318	96.141,318	100.133,739

(continued)

Table 2.1 (continued)

CPA08/TIME	2009	2010	2011	2012	2013	2014	2015	2016
Wood and of products of wood and cork, except furniture; articles of straw and plaiting materials	22.118,732	23.578,096	20.909,494	20.477,622	22.096,348	20.456,91	22.581,359	20.702,592
Paper and paper products	22.480,893	21.865,07	21.915,109	25.717,311	25.065,943	25.521,794	30.392,517	30.436,955
Printing and recording services	4.781,105	4.243,494	3.952,707	3.388,825	3.431,419	3.725,877	4.385,756	5.219,973
Coke and refined petroleum products	59.505,41	58.228,144	50.751,1	47.978,79	53.936,777	50.431,967	61.134,651	61.115,829
Chemicals and chemical products	57.392,335	62.940,772	60.771,194	71.826,369	81.834,247	103.879,323	111.639,893	99.943,131
Basic pharmaceutical products and pharmaceutical preparations	39.299,372	36.651,835	34.416,113	33.995,372	39.154,024	41.708,505	48.712,745	49.091,717
Rubber and plastic products	19.216,414	18.715,751	21.347,142	17.878,434	18.674,975	22.446,315	23.014,986	28.303,799

(continued)

Table 2.1 (continued)

CPA08/TIME	2009	2010	2011	2012	2013	2014	2015	2016
Other non-metallic mineral products	103.780,258	94.574,264	89.979,345	83.962,184	83.329,008	62.925,847	77.344,742	54.786,501
Basic metals	27.892,578	35.038,133	46.827,066	4.065,378	17.272,769	27.830,296	21.393,241	32.489,063
Fabricated metal products, except machinery and equipment	63.208,273	68.216,113	69.151,226	65.891,356	67.353,517	74.795,049	76.455,522	90.028,301
Computer, electronic and optical products	89.016,212	92.092,95	81.134,705	69.288,812	72.695,472	77.515,796	83.968,515	88.239,563
Electrical equipment	70.713,698	75.494,44	71.726,64	66.597,756	64.617,031	68.290,103	69.821,3	76.622,122
Machinery and equipment n.e.c.	112.932,733	122.135,802	129.712,761	115.732,262	118.247,876	129.017,639	133.671,504	154.685,09
Motor vehicles, trailers and semi-trailers	174.244,147	186.447,145	184.399,182	168.688,606	164.656,743	180.942,971	205.310,31	235.725,488
Other transport equipment	46.062,799	48.992,846	47.057,619	41.711,556	37.919,956	40.081,084	40.679,599	42.284,611
Furniture and other manufactured goods	123.774,137	126.153,994	118.604,914	110.940,552	115.730,302	127.688,943	130.861,344	145.818,446

(continued)

Table 2.1 (continued)

CPA08/TIME	2009	2010	2011	2012	2013	2014	2015	2016
Repair and installation services of machinery and equipment	40.006,802	41.452,875	40.365,823	36.807,621	38.132,39	40.326,497	41.318,312	44.533,537
Electricity, gas, steam and air conditioning	302.650,112	301.117,773	295.187,433	299.437,177	288.935,412	267.562,228	269.369,079	254.964,201
Natural water; water treatment and supply services	17.205,331	16.896,087	16.651,561	18.437,213	18.592,641	15.092,86	14.680,472	16.755,295
Sewerage services; sewage sludge; waste collection, treatment and disposal services; materials recovery services; remediation services and other waste management	73.342,189	65.743,236	71.201,697	60.010,187	55.065,134	53.930,035	55.190,401	50.907,794

(continued)

Table 2.1 (continued)

CPA08/TIME	2009	2010	2011	2012	2013	2014	2015	2016
Constructions and construction works	2.370.463,4	2.390.434,57	2.559.409,385	2.367.603,701	2.144.972,881	2.001.511,973	1.951.378,347	1.807.544,387
Wholesale and retail trade and repair services of motor vehicles and motorcycles	57.093,289	53.502,852	55.852,814	50.571,2	51.785,453	56.176,418	61.170,171	65.591,923
Wholesale trade services, except of motor vehicles and motorcycles	165.419,267	149.443,597	168.602,393	145.767,351	162.018,014	176.027,66	175.606,794	178.332,852
Retail trade services, except of motor vehicles and motorcycles	185.946,888	171.922,93	168.435,41	136.146,642	128.576,524	136.144,037	134.577,756	132.206,009
Land transport services and transport services via pipelines	182.235,822	171.209,915	176.538,611	167.992,716	171.785,354	186.936,139	202.207,965	211.065,892
Water transport services	5.654,728	5.392,257	7.077,705	6.195,403	7.423,401	7.168,647	7.072,508	6.488,345

(continued)

Table 2.1 (continued)

CPA08/TIME	2009	2010	2011	2012	2013	2014	2015	2016
Air transport services	17.415,064	17.360,162	16.733,471	17.696,43	17.169,776	18.455,654	18.858,592	19.673,607
Warehousing and support services for transportation	20.160,298	19.180,333	19.399,043	19.433,888	18.641,795	20.602,686	21.707,976	21.909,72
Postal and courier services	1.825,018	1.712,948	1.677,03	1.672,619	1.570,13	1.592,284	1.718,012	1.736,858
Accommodation and food services	273.171,292	258.063,665	254.109,604	229.756,571	233.730,766	249.070,465	245.553,49	253.193,443
Publishing services	24.096,02	22.637,473	18.406,047	16.451,758	16.575,952	18.477,866	20.473,945	22.593,7
Motion picture, video and television programme production services, sound recording and music publishing; programming and broadcasting services	19.179,864	18.072,103	16.003,427	13.689,591	14.016,993	14.065,104	15.113,712	15.374,958

(continued)

Table 2.1 (continued)

CPA08/TIME	2009	2010	2011	2012	2013	2014	2015	2016
Telecommunications services	38.935,322	35.966,529	34.645,487	32.447,975	31.353,316	30.632,279	31.180,051	31.442,68
Computer programming, consultancy and related services; Information services	28.064,129	28.245,118	28.280,829	26.309,877	28.501,499	29.806,587	31.986,828	33.906,977
Financial services, except insurance and pension funding	17.416,655	16.264,521	16.224,964	14.268,707	13.726,035	14.536,912	15.322,15	15.191,955
Insurance, reinsurance and pension funding services, except compulsory social security	28.130,872	25.944,95	25.563,62	25.102,428	24.908,368	25.203,05	26.415,788	26.039,397
Services auxiliary to financial services and insurance services	2.016,752	1.907,919	1.737,546	1.811,212	1.795,97	1.907,115	1.976,222	1.990,103
Real estate services	290.622,825	275.610,445	283.336,96	258.770,91	260.462,853	249.117,293	245.867,052	235.353,125

(continued)

Table 2.1 (continued)

CPA08/TIME	2009	2010	2011	2012	2013	2014	2015	2016
Legal and accounting services; services of head offices; management consultancy services	6.791,992	6.916,675	7.489,21	6.993,093	7.418,676	8.040,248	8.661,745	8.977,041
Architectural and engineering services; technical testing and analysis services	18.044,1	18.505,57	19.359,653	17.022,054	16.981,513	17.768,421	19.215,85	19.988,223
Scientific research and development services	79.375,24	80.915,62	83.623,747	75.358,417	80.116,906	86.758,669	89.928,784	96.829,361
Advertising and market research services	130,72	124,92	3,629	87,561	190,409	342,438	268,023	417,949

(continued)

Table 2.1 (continued)

CPA08/TIME	2009	2010	2011	2012	2013	2014	2015	2016
Other professional, scientific and technical services and veterinary services	4.470,505	4.319,748	4.396,803	4.055,861	4.176,535	4.677,96	5.038,719	5.108,951
Rental and leasing services	8.127,927	7.601,789	7.518,413	7.775,194	8.280,355	8.311,548	9.043,07	9.358,035
Employment services	411,476	410,685	304,313	271,534	295,547	364,131	381,578	344,963
Travel agency, tour operator and other reservation services and related services	12.708,437	12.083,008	11.586,303	10.176,86	10.236,886	10.550,658	10.887,42	11.561,252
Security and investigation services; services to buildings and landscape; office administrative, office support and other business support services	15.537,551	12.768,215	12.584,564	9.720,099	10.559,496	12.925,18	15.091,844	15.708,409

(continued)

Table 2.1 (continued)

CPA08/TIME	2009	2010	2011	2012	2013	2014	2015	2016
Public administration and defence services; compulsory social security services	264.937,186	228.237,445	218.471,03	188.194,603	202.084,995	240.038,882	260.117,292	255.749,767
Education services	84.835,476	81.503,354	77.027,551	73.893,799	71.874,878	74.741,722	75.417,371	72.849,081
Human health services	164.684,251	157.682,222	154.873,213	144.778,441	145.406,948	158.040,265	161.913,391	166.607,274
Residential care services; social work services without accommodation	63.925,237	61.040,897	61.143,237	54.977,431	55.894,006	59.501,31	60.716,114	59.024,43
Creative, arts, entertainment, library, archive, museum, other cultural services; gambling and betting services	31.103,019	29.840,539	28.802,184	26.442,062	26.800,812	28.300,493	28.989,423	29.218,482

(continued)

Table 2.1 (continued)

CPA08/TIME	2009	2010	2011	2012	2013	2014	2015	2016
Sporting services and amusement and recreation services	25.376,953	23.488,153	22.391,43	21.816,888	21.816,028	22.758,607	22.350,153	22.145,422
Services furnished by membership organisations	22.283,98	21.371,494	20.315,531	19.330	18.955,453	19.280,995	19.818,585	19.648,83
Repair services of computers and personal and household goods	5.122,132	4.902,424	4.437,504	4.304,471	4.160,298	4.450,934	4.666,611	4.743,976
Other personal services	23.341,318	22.774,198	22.892,846	23.783,786	23.633,454	23.823,364	24.154,093	23.134,776
Services of households as employers; undifferentiated goods and services produced by households	673,523	616,349	568,446	420,672	444	415,089	426,265	423,966

Source (Eurostat)

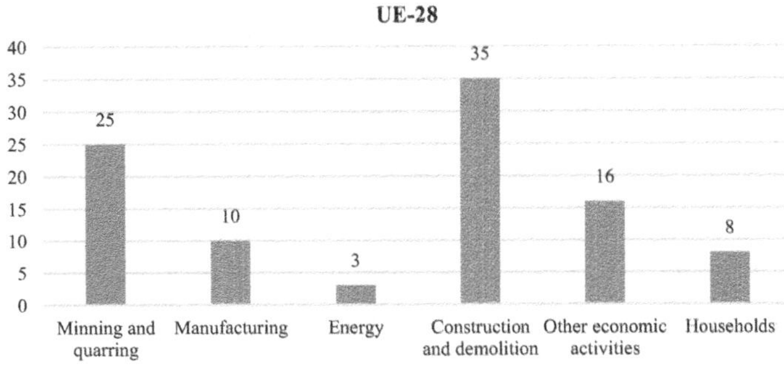

Fig. 2.2 Waste generation by economic activities and households, 2016. Data from Eurostat (env_wasgen)

more than the 30% of the total waste generated (Fig. 2.2). It consists of a mineral fraction (bricks, tiles, concrete, etc.), which is relatively heavy and easily recyclable (but of low value) and also contains materials with positive market value (metals) or potential value if collected separately in clean fractions (such as plastics) [12]. As the EC study "Resource Efficient Use of Mixed Wastes. Improving management of construction and demolition waste" [13] highlights: "*European statistics do not permit a break-down of CDW by activity (construction, refurbishment, demolition), or by subsector (public works, buildings), and too few information could be retrieved at MS level to perform an estimate. However, the data available provides a break-down of the material included in CDW: the majority of CDW generated in Member States is composed of mineral waste from construction and demolition*".

The EC study has also shown that a deeper analysis regarding the breakdown by material type is possible only for some selected Member States (Germany, Denmark, Estonia, Croatia, Hungary, Luxembourg, Portugal, Slovakia). A breakdown of the materials in CDW for these Member States (MS) has been based on the EWC top level categories (17 01, 17 02 etc.).[2] In Fig. 2.3 a percentage of the total CDW generated (excluding soils, 17 05).

The breakdown of materials, even if considered for a limited number of Member States, highlights a high degree of variability and that, for most MS, the major waste types are concrete, bricks, tiles and ceramics. These data are consistent with the fact

be understood as corresponding to the types of waste included in Chapter 17 of the list of waste established by Decision 2014/955/EU in the version in force on 4 July 2018" [11].

[2]Eurostat waste statistics are available at the three digit level of the European Waste Catalogue (EWC_Stat), which is a mainly material-based classification of waste. The EWC_Stat codes do not include information regarding the waste generating activity, except for some EWC_Stat codes such as W121, which specifically refers to mineral waste from construction and demolition. Yet Eurostat data on generation are reported by the countries according to different NACE activities. This means that EWC-Stat data does not allow to identify with certainty whether waste originates from construction and demolition activities, except for waste generated and reported under the construction sector (Nace F). It would however be an underestimation to only consider wastes reported in Nace F, as Member States are known to also report generated CDW in other sectors than the construction sector [13].

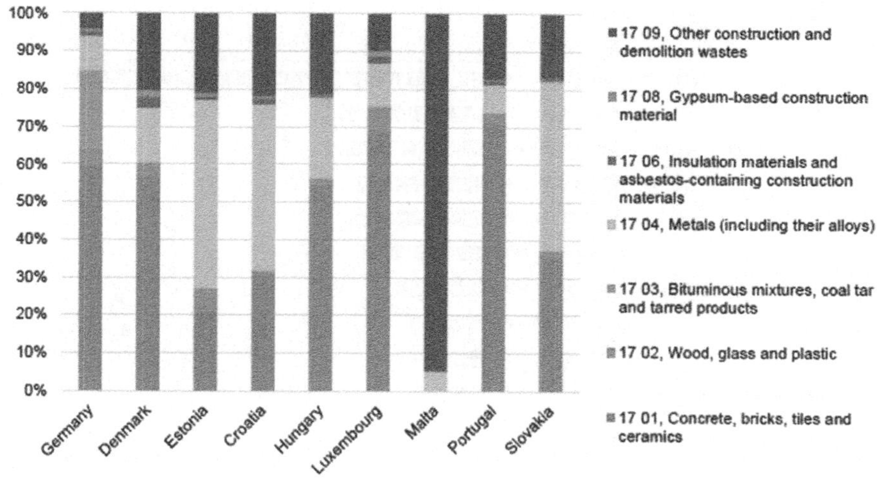

Fig. 2.3 Breakdown of CDW generated by material type. *Source* EC [13]

that the major fraction of CDW generated in Member States is composed of inert waste. "*But it seems unlikely that any MS would have more metal-based waste being generated than concrete and brick based waste, as this seems to be the case in Estonia, Croatia and Slovakia. In this regard, it is to be noted that the figures shown here are higher than those reported in official data*" [13]. Probably, large amounts of 'inert' CDW are not being reported, perhaps due to very high levels of on-site reuse. If this is happening, the overall waste generation data would be significantly underestimating the actual amount of CDW that is generated. The report emphasizes the variability of data quality, that makes difficult to carry out a quantitative comparison of MS performances and have confidence in the trends.

Being aware of this problem, it is possible to focus the attention also on another issue, connected with CDW, related to the waste stream, made up of a mix of different materials including inert, non-inert non-hazardous and hazardous waste.

A survey [14], developed in 2009, has estimated the transboundary movements of waste considering 21 European countries. The survey has high-lighted that the list, based on the code categorization ELW (European List of Waste) [15], of the 10 hazardous waste types (Fig. 2.4), exported in greatest quantities, is dominated by construction and demolition waste (codes starting with 17), including contaminated soil, contaminated wood and asbestos. "It is perhaps surprising that large amounts of soil and asbestos waste are shipped across borders since they are usually heavy and therefore expensive to transport. Contaminated soil is exported mainly because of a lack of treatment capacity in certain countries [16]. Soil is cleaned or placed in a landfill, asbestos waste is usually landfilled and contaminated wood is incinerated. In countries with strong policies on renewable energy power, plants fed with biomass need large amounts of fuels, attracting wood waste from abroad [16]" [14].

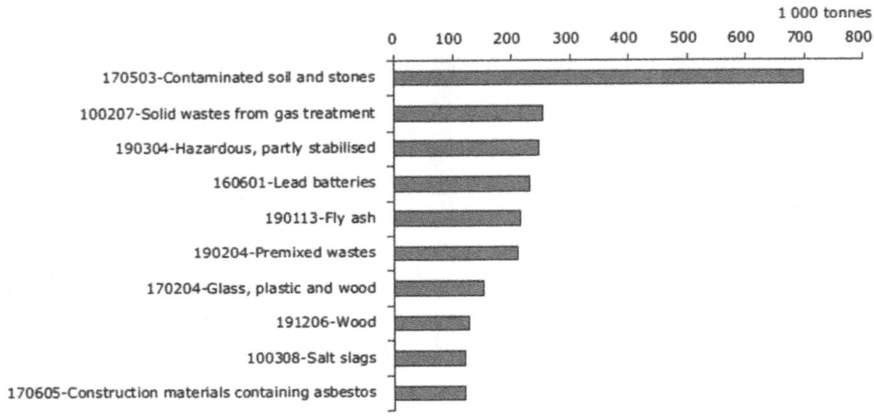

Fig. 2.4 Top 10 hazardous waste types by amounts exported according to ELW code categorisation, 2007. Data from 16 countries. *Source* EEA [14]

2.2 Key Actions for the Reduction of the General Environmental Impacts of the Construction Sector Related to Waste Production

The Waste Framework Directive sets a 2020 target of 70% preparation for re-use, recycling and other material recovery for this waste stream. Member States are variously reaching this target as reported in Table 2.2.

Surely, in order to improve the resource efficiency and to reduce the general environmental impacts of the construction sector, related to waste production, many and integrated key actions may be identified [17]. These actions deal with political, organizational and technical aspects, may be applied at various scales (from the micro scale of the material to the macro scale of Real estate) [18] and involve many different stakeholders (policy makers, standardizers, large, medium-sized and small manufacturing enterprises, owners, designers, State managers, construction firms, companies specialized in activities of demolition, users, etc.). These actions, aiming both to improve the use of resources and to reduce the amount of waste, may orient to:

- use building products characterized by a low content of materials and embodied energy;
- use building products containing secondary raw materials, obtained through recycling processes (both open and closed loop[3]) [20] of scraps coming both from

[3]*"The international standard on life cycle assessment (BS EN ISO 14044, 2006) distinguishes between two different types of recycling process, "closed-loop" and "open-loop". Closed-loop recycling occurs when either, (i) the product is recycled at the end of its life into the same product system (...) or (ii) the product is recycled into a different product system but the materials undergo no change in inherent properties (...). In both cases (i) and (ii), the material resulting from the recycling process displaces virgin production in a 1:1 ratio, and the environmental benefits allocated*

Table 2.2 Recovery rate of construction and demolition waste % of construction and demolition mineral waste recycled

Geo\time	2010	2012	2014	2016
EU (28 countries)	79	86	89	90
Euro area (19 countries)	:	:	:	:
Belgium	17	18	32	95
Bulgaria	62	:	:	90
Czechia	91	91	90	92
Denmark	:	91	92	90
Germany	95	94	:	:
Estonia	96	96	98	97
Ireland	97	100	100	96
Greece	0	0	0	:
Spain	65	84	70	79
France	66	66	71	72
Croatia	2	51	69	76
Italy	97	97	97	98
Cyprus	0	60	38	57
Latvia	:	:	92	98
Lithuania	73	88	92	97
Luxembourg	98	99	98	100
Hungary	61	75	86	99
Malta	16	100	100	100
Netherlands	100	100	100	100
Austria	92	92	94	88
Poland	93	92	96	91
Portugal	:	:	:	97
Romania	47	67	65	85
Slovenia	94	92	98	98
Slovakia	:	:	54	54
Finland	5	12	83	87
Sweden	78	81	55	61
United Kingdom	96	96	96	96
Iceland	75	100	86	99
Liechtenstein	:	:	:	:

Data from Eurostat, dataset (cei_wm040)

manufacturing sectors (cross sectoral recycling) and the construction sector (sectoral recycling), both from the pre-consumer and the post-consumer phase (see Chap. 3);

- use building products obtained from pre-consumer by-products coming both from manufacturing sectors and the construction sector (see Chap. 3);
- extend the lifespan of the buildings elements through improvements in the design of the products based on approaches such as smart design for adaptability,[4] design for maintainability [22] and design for repair [18, 21, 23];
- extend the lifespan of the building elements through the improvement of the durability of the products [24, 25] and the application of practices of planned maintenance [26, 27];
- prevent waste through smart design for disassembly and deconstruction [28, 29];
- reduce waste generation by supporting decision making in the design phase[5];

to recycling reflect this. Conversely, in open-loop recycling the recycled material undergoes a change in inherent properties (usually a loss of quality). The environmental benefits associated with the recycling process are then not those associated with displaced virgin production of the original material and must be accounted for in some other way e.g. displacement of other material production" [19].

[4]"*Adaptability is a design characteristic which embodies spatial, structural, and service strategies, allowing the physical artefact a level of malleability to respond to users' needs over time, enable improving the building's lifespan. Thus, buildings' adaptability can be said as a cornerstone of sustainability. The assumption is that the building is more sustainable if its transformation capacity is higher. The transformation capacity of buildings can improve their flexibility, as it allows building to change, and adapt to new requirements. If a building can be rearranged to embrace change, modifying internal spaces for various uses, its flexibility increases*" [21].

[5]As an example the Designing out Waste Tool for Buildings (DoWT-B) is a decision making tool, that can help the user identify opportunities to prevent waste through the application of design solutions, and calculate the combined impact of the chosen design solutions. The DoWT-B has been designed to be used at RIBA stages A, B or C. and can be used to facilitate designing out waste in buildings projects by providing outputs based on limited input data. The Tool can help the design team to:

- generate indicative waste forecasts;
- identify the building elements that are likely to produce the most waste;
- identify opportunities to design out waste and calculate the combined impact of selected design solutions;
- assess the impact of the chosen design solutions on the value of waste, waste to land-fill, disposal cost and CO_2 from materials extraction [30].

DoWT-B is a freely accessible online resource, available at www.wrap.org.uk/nwtool.

- improve practices such as reuse,[6] recycling,[7] reconditioning[8] and remanufacturing[9] of building components and products (Fig. 2.5) through design for modularity,[10] disassembly, deconstruction and adaptability [33];

[6]"*Re-use means any operation by which products or components that are not waste are used again for the same purpose for which they were conceived. Preparing for re-use' means checking, cleaning or repairing recovery operations, by which products or components of products that have become waste are prepared so that they can be re-used without any other pre-processing*" (Dir. 98/2008).

[7]"*Recycling means any recovery operation by which waste materials are reprocessed into products, materials or substances whether for the original or other purposes. It includes the reprocessing of organic material but does not include energy recovery and the reprocessing into materials that are to be used as fuels or for backfilling operations*" (Dir. 98/2008).

[8]Recondition: "*return a used product to a satisfactory working condition by rebuilding or repairing major components that are close to failure, even where there are no reported or apparent faults in those components. With respect to reconditioning: manufacturing effort involves the replacement of worn or broken parts, generally less extensive than required to remanufacture, but more than necessary for repair; performance after reconditioning is expected to perform its intended role but the overall performance is likely to be inferior to that of the original model; any subsequent warranty is generally less than new or a remanufactured product but the warranty is likely to cover the whole product (unlike repair); reconditioned products do not require a warranty equivalent to that of a newly manufactured equivalent*" [31].

[9]"*Remanufacture return a used product to at least its original performance with a warranty that is equivalent or better than that of the newly manufactured product. From a customer viewpoint, the re-manufactured product can be considered to be the same as the new product. With respect to remanufacture: manufacturing effort involves dismantling the product, the restoration and replacement of components and testing of the individual parts and whole product to ensure that it is within its original design specifications; performance after remanufacture is expected to be at least to the original performance specification; any subsequent warranty is generally at least equal to that of new product. This assumes that remanufacture applies to like-for-like products*" [31].

[10]Modular assembly: "*Assembly made up of a series of components, which may be exchanged for others, using the same pre-defined interfaces*" [31]. "*Modular design is widely applied in the construction sector, where buildings are produced in modules at a factory and subsequently assembled on site. Modular construction contributes to circularity in several ways. First, waste is more readily reduced in a controlled environment such as a factory, where practices such as recycling of materials, controlling inventory and protecting building materials are more easily implemented than on an open construction site that is more prone to external disturbance. Modular construction typically*

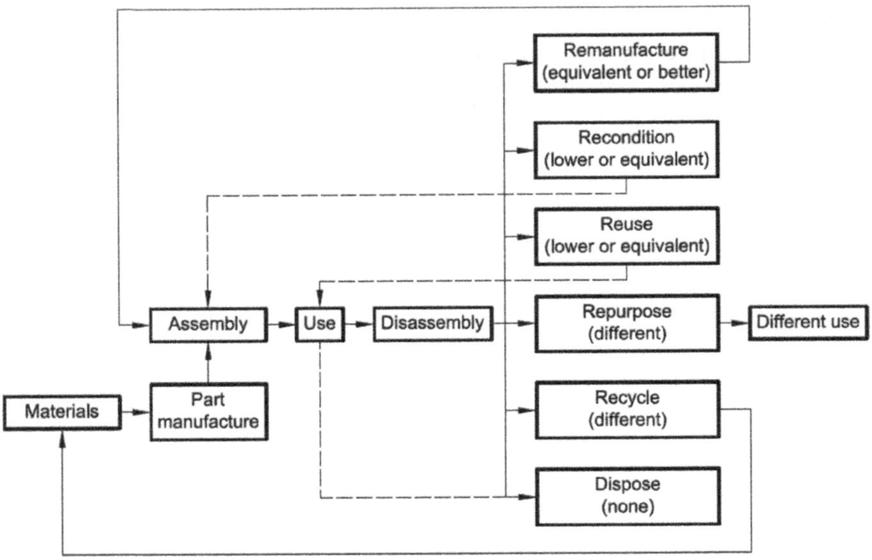

Fig. 2.5 Overview of design for manufacture, assembly, disassembly and end-of-life processing (MADE). *Source* BS 8887: 2009 [31]

– improve the reuse and recycling of building materials and products through technics and procedures, especially pre-demolition waste audits[11] [36], pursuing the optimization of the disassembly and demolition activities [37, 38];

involves less transport of materials and staff, contributing to fewer emissions (…). Moreover, modular buildings can be disassembled and the modules relocated or refurbished for reuse, reducing the demand for raw materials and minimising the amount of energy expended in creating a building to meet the new need. The potential reusability of detachable components raises the resale value of building parts that can be replaced, recycled or moved according to need. However, this might be challenging due to the generally long lifetime of buildings, leading to the probability that modules will be outdated by the time they become available for reuse. Finally, modular buildings make the repair or modification of materials or parts possible without destroying buildings' basic structure" [32].

[11]"*A waste audit before demolition or renovation of buildings and infrastructures is a specific task within the project planning. It is necessary to understand the type and amount of elements and materials that will be deconstructed and/or demolished, and to issue recommendations on their further handling. An assessment of the viable recovery routes for materials can also be given (including reuse and the potential reuse value, recycling on- and offsite and the associated cost savings and energy recovery) (…). Performing a waste audit presents a series of advantages—both economic and environmental—providing important added value to the whole project:*

- *Waste audits are the first step towards recycling*
- *Waste audits promote fair competition amongst contractors*
- *Waste audits increase awareness and ease traceability processes. It is of major importance to know the materials that will be set free; especially the hazardous ones to avoid unexpected costs during the works*
- *Environmental and technical quality of materials can be steered.*

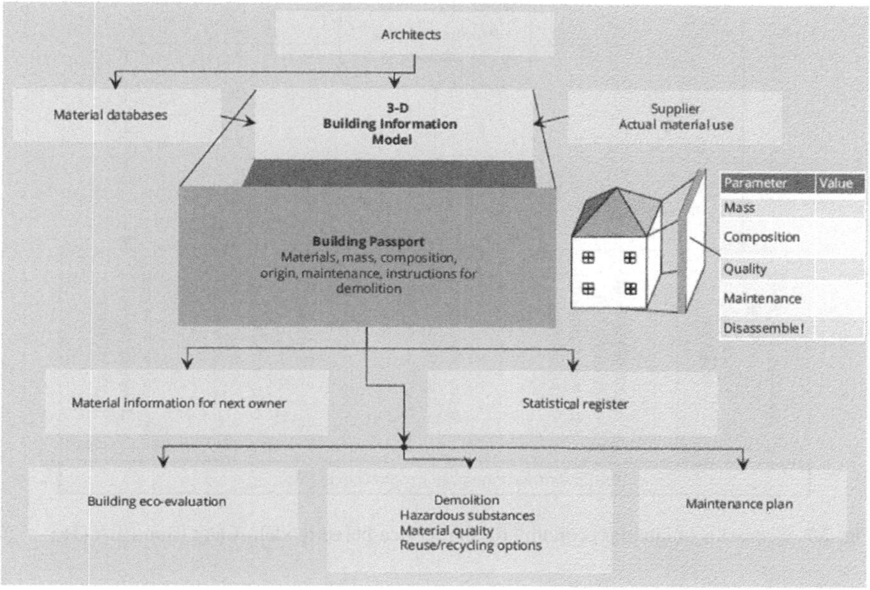

Fig. 2.6 Scheme of a building material information system: the Austrian case of the building passport. *Source* EEA [39]

- develop standards and parameters for assessing the quality of reused, recycled, remanufactured materials and products;
- reduce offcuts and surplus materials within the construction site through a smart planning of the construction phase [33];
- share offcuts and surplus materials within the construction site through planned practices of exchange with nearby construction sites [33];
- boost[12] the transitioning from a product sales model to a service-based model (Figs. 2.6 and 2.7) [39, 32];

Sources: EU, Guidelines for the waste audits before demolition and renovation works of buildings, EU Construction and Demolition Waste Management [11] and the Kretsloppsrådet report [34]. See also the DG GROW study [35], that has developed a methodology guidance for pre-demolition audit. The guidelines aim at establishing the basic principles of waste auditing, identification, reporting and recommendations for the further treatment (EC "Technical and Economic Study with regard to the Development of Specific Tools and/or Guidelines for Assessment of Construction and Demolition Waste Streams prior to Demolition or Renovation of Buildings and Infrastructures". Final report of EU Specific Contract 30-CE-0751644/00-00 – SI2.720069 (2016).

[12]The transition towards a service-based business model must face various challenges. It needs the acquisition of new references and practices, the definition of new financing mechanisms and of effective systems for collecting and possible reusing the products at the end of a service agreement. Moreover production planning and logistics need predictions about the time the various products in a service will reach the end of their technical or useful life.

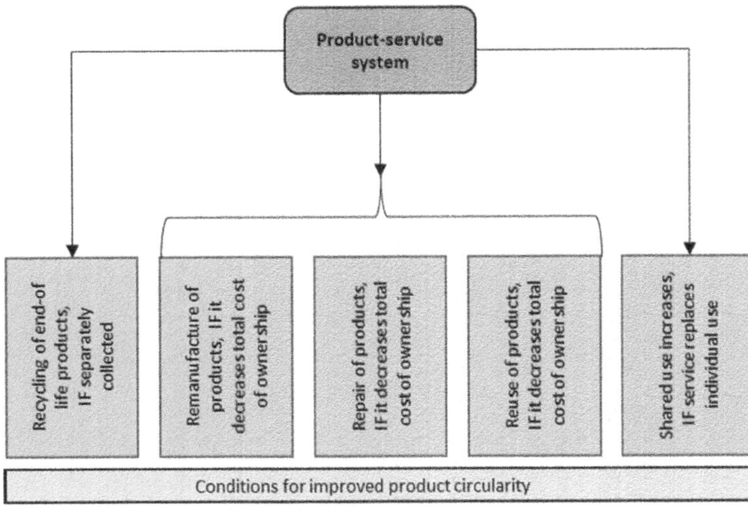

Fig. 2.7 Scenarios of circular economy for the service-based model. Adapt from EEA [18]

- boost the market of reuse and recycle of building materials and products through the improvement of information[13] and the development of organizational supports (e.g. exchange platforms) (See Chap. 5);
- promote the maintenance [40] and renovation of real Estate instead of new construction of buildings;
- support the creation of networks[14] for the exchange of reused building products;

[13]In order to pursue waste prevention in the construction and infrastructure sectors, Austria has introduced a 'building passport' as part of a building material information system. The passport is part of the building's documentation throughout its life cycle (Fig. 2.6). The passport aims to connect architects, suppliers and statistical registers to enable the selective demolition of buildings, the reuse and high-quality recycling of building materials, and the prevention of waste by extending the lifespan of buildings through maintenance plans. The passport includes all the in-formation useful for the waste-light operation of buildings and records all building activities, systems and items. Following several pilot projects, the Austrian waste prevention programme is working to standardise the passport and increase its use [39].

[14]Various charitable reuse organisations in Europe help households in need by supporting them in obtaining used furniture and building components. As an example in the United Kingdom the Furniture Re-use Network is a network that promotes the reuse, recycling and refurbishment of used furniture, supports and assists people in obtaining reused furniture at affordable prices and in providing training and work placement opportunities. See: FRN (2017), Furniture Re-use Network (http://www.frn.org.uk).

- support the establishment of centers and markets for used building materials,[15] that can be managed both by private operators and by municipalities [18];
- develop guidelines for public procurement on new constructions, renovations and infrastructure construction and management, oriented to material efficiency according to the circular economy strategies [18].

2.3 Barriers and Drivers for a Market of a CDW Recycling Market

According to the Frost and Sullivan report "*Construction and Demolition Recycling*",[16] the Construction and demolition waste (CDW) recycling market in Europe had a value of around EUR 13.8 billion in 2013 and it is expected to grow to EUR 17.6 billion by 2020 in parallel with a likely increase in the volume of CDW.

But, notwithstanding the feasibility of the many actions, above described, the actual growth and maturation of a market of CDW depend on the overcoming of some obstacles.

Many, various possible, often interacting, barriers may prevent the full development of this kind of market:

- wastes coming from the buildings are heterogeneous mixtures (clay brick, mineral bounded building materials, like concrete, calcium silicate unit, aerated concrete and lightweight concrete, mortar, plaster, insulation material, wood, plastic etc.). "*Even after processing—such as presorting by hand, single or double-stage crushing, separation of reinforcement by over belt magnetic separators, air sifter for the separation of light components, jigging machines and sieving—the recycled product often remains heterogeneous. In the building industry the amount of composite building materials is increasing nowadays, leading to a more difficult separation*" [41];

[15]As an example [18], Bauteilbörsen is a Swiss business to business (B2B) non-profit company, established in 1995. It offers a wide range of second-hand building components in good condition, such as: parquet flooring, natural stone, partitions, bricks, doors, windows, glass, stairs, bath tubs, ba-sins, fittings, washing machines, water heaters, mirrors, tiled stoves, fitted kitchens, ovens, wash basins and fridges. The company provides dismantling and demolition services, the sale of used elements, the registration in a country-wide online database, and consultancy on the components' reuse according to the needs and wishes of the customers. It is a 'virtual warehouse' where the available items can be registered. In Germany, Bauteilbörse Bremen collects used building components and structural elements during demolition or reconstruction processes, such as: floors, doors, windows, stairs, kitchens, etc. Bauteilbörse Bremen mainly represents the communication portal and organization tool between those who offer and those who need building components and elements, that can be reused. Bauteilbörse Bremen has a warehouse, but where it is possible, the exchange between both parties is realised on-site. See: Bauteilbörse Basel [30] (http://www.btbbasel.ch/ueber-uns.html); (http://www.bauteilboersebremen.de).

[16]http://www.cdrecycler.com/article/frost-sullivan-report-construction-demolition-market/.

- buildings are complex entities composed of many different long-life components. "*While most circular economy thinking is focused at the individual product level, buildings are compositions of numerous products with different lifespans, which may be altered in a number of ways, from routine maintenance to structural adaptation*". Besides, "*over a building's lifespan, developments in materials, regulations, construction methods, and societal needs can create functional obsolescence of deconstructable buildings and reusable components; unexpected damage or wear can reduce integrity*" [42]. These factors may represent an obstacle in the assessment of the actual characteristics of the products and in the feasibility of circular processes;
- "*low prices of raw materials, combined with limitations in the market for secondary materials pose a challenge for CDW reuse and recycling, as the quantities produced might not be absorbed by the market*"[17];
- in a vicious cycle, a weak demand for some recycled resources limits the investment in innovation, the lack of investment in supply can leave potential buyers uncertain about high-quality supplies, weakening demand for those orienting investments [1];
- risk of an unbalanced market with oversupply of some recycled materials and shortages of others. For instance, large amount of recycled aggregates are mainly used in low-grade applications (e.g. subfoundations) [43] and this market is getting increasingly saturated.[18] At this regard, the EC report" *Resource Efficient Use of Mixed Wastes—Improving management of construction and demolition waste*" [13] analyses of the amount of recycled aggregate produced in some MS (Fig. 2.8), showing that this quantity is relatively small compared to the total aggregate demand: as an average, 9% of the demand for aggregates is covered with recycled aggregates. "*This suggests that the market could absorb increasing amounts of recycled aggregates in most MS. However, it should be noted that the availability of recycled aggregate is likely to vary over time and regionally with some areas having a shortage of materials while other areas may have a surplus as*

[17]In Sweden for example, the low cost of building materials, compared to the expensive labour cost, causes extra volumes of materials to be ordered to prevent construction delays. Such delays would translate into higher costs for contactors, having to pay for workers waiting for materials to be delivered. The unused materials are then often sent to landfills [37].

[18]For instance, in the Netherlands, the majority of CDW (more than 95%) is currently recycled. A large amount of most materials used in buildings (the residential and non-residential building sector) after demolition and recycling is used in civil engineering, often as a road base material or as filler material to raise the level of industrial estates. In civil engineering, more than 50% of the materials used (excluding earth-moving) consists of recycled materials, however, in the building/construction sector hardly any secondary material is used, no more than 3–4%. There is a gradual saturation of secondary materials in civil works. The demand is further reduced because fewer residential areas and industrial estates are being built on greenfield sites, and the Provinces, district water boards and Rijkswaterstaat are shifting away from the building of new infrastructure towards maintenance of existing assets. The risk is now a long expected surplus of recycled aggregates [41].

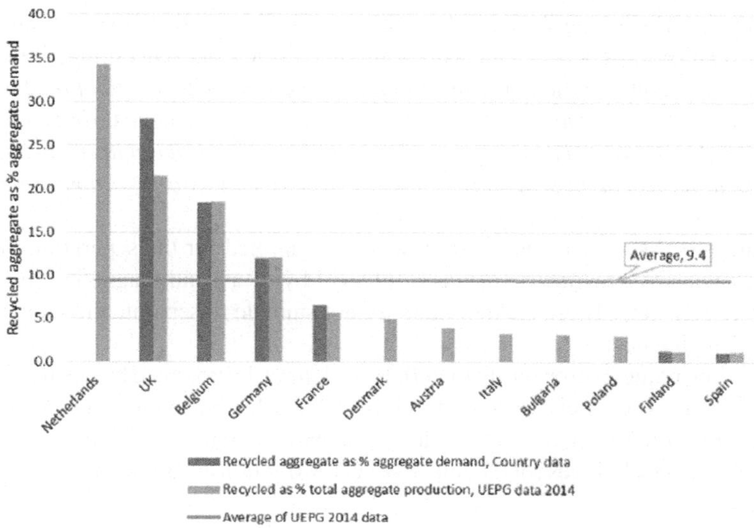

Fig. 2.8 Overview of the proportion of national aggregate demand met by recycled aggregates across MS. *Source* EC [13]

highlighted in a German study[19] *of long-term potentials of high-grade recycling of mineral construction waste*" [13];

- difficulties in quantifying the actual demand and in planning the availability of materials and products to treated, owing to the uncertainty and high variability of the characteristics of the waste;
- most buildings, built from the mid-twentieth century onwards, are not constructed in a way that gives high potential for components recovery and this makes difficult to consider many existing buildings as a direct source of recyclable and reusable components[20] [42];
- CDW are mainly heavy and bulky materials, often need to be moved over long distances to reach their treatment plants and subsequently their markets. These aspects increase significantly the costs and the environmental impacts;

[19]See: Determining Resource Conservation Potentials in the Recovery of Construction Waste and Formulating Recommendations on Their Use Schiller et al., Environmental Research of the Federal Ministry of the Environment, Nature Conservation and Nuclear Safety, 2010. https://www.umweltbundesamt.de/sites/default/files/medien/461/publikationen/texte_56_2010_kurzfassung_e1_0.pdf.

[20]"*While most circular economy thinking is focused at the individual product level, buildings are compositions of numerous products with different lifespans, which may be altered in a number of ways, from routine maintenance to structural adaptation. These factors mean that buildings are complex entities, and represent a different challenge to that of individual products. Circular economy models are yet to reach widespread application beyond specific components that are easily removed and need frequent replacement, such as lighting and carpet tiles. Such ongoing contractual arrangements between building owner and manufacturer have great benefits in forcing engagement with future end-of-life. Their suitability to longer-lived components is less clear, given the likelihood of manufacturers ceasing trading before the circle closes*" [43].

- the organization of the selective demolition works[21] (variations in the usual skills, in the relationships between the work teams and in the work scheduling), if not well managed, might lengthen the time and increase costs. "*The length of time needed to deconstruct can be unappealing where extra costs are incurred through having a building (such as local property taxes) or loss of revenue on a replacement building owing to an extended scheduling of works. There can also be a time constraint linked to planning permission expiration*" [33];
- sorting processes have been until now mainly applied for the separation of light components by gravimetric means or of steel by over-band magnetic separators, respectively [45]. These technologies are not suitable to separate the incidentally mixed aggregates;
- up-cycling requests high purity materials, which can be obtained by a better sorting at the source, (i.e. selective demolition) [44], but selective manual deconstruction, separation and storage of waste are labor-intensive activities in the face of a lack of cost efficient techniques for the selective deconstruction and removal of "complex" materials;
- fractions of waste are often unknown in their characteristics and composition (with the possibility of the presence of hazardous waste [46–48]). This implies high levels of uncertainty generating health and safety risks to be carefully and specifically managed [49];
- the need for dedicated storage areas for separated waste may impede the works, delay the execution time and, consequently, increase the costs;
- clients and designers may be prejudiced towards the products deriving from reuse, recycling, remanufacturing, etc. owing to the lack of a long-established practices of certification[22] of tested performances and warranties;

---

[21]"*The extra costs for selective demolition (including labor costs, extra containers and transport) need to be countered by lower disposal costs (or a higher positive value) for the produced fractions. Better valorization routes and/or improved logistics are needed for certain fractions (e.g. insulation materials, gypsum)*" [44].

[22]The EU Construction & Demolition Waste Management Protocol [49] identifies three possible levels of actions to validate the quality of recycled and reused materials and components:

- make use of existing European product standards. "*The Construction Products Regulation (EU/305/2011, CPR) lays down harmonised rules for the marketing of con-struction products and provides tools to assess the performance of construction prod-ucts. Construction products that are covered by Harmonised European Standards (hENs) need a Declaration of Performance (DoP) and have to be CE-marked to in-crease transparency*";
- make use of European Technical Assessments. "*Products that are not (fully) covered by hENs can still be CE-marked with the use of European Technical Assessments (ETA) issued according to European Assessment Documents (EAD). The ETA document provides information about the performance of a construction product, to be declared in relation to its essential characteristics. This voluntary tool enables manufacturers to place recycled or re-used products on the EU market and allows them to declare specific information about the performance of their products. There are al-ready examples of using these tools for processed demolition waste, mainly for recycled aggregates*";
- make use of Quality Assurance schemes. "*In several Member States there are Quality Assurance schemes in place for specific products, like recycled aggregates. Such schemes often contain requirements concerning waste acceptance and environmental issues. When working with such*

- poorly informed customers may have bias considering as lower quality products the ones deriving from recycling, reuse, remanufacturing processes.

Various drivers can help to overcome these barriers and support the expected growth of the "re-materials" and products market [37], such as:

- optimized advanced technologies for demolition and sorting[23] of waste[24];
- promotion of selective demolition[25] by the authorities through specifications within public procurements;
- regulatory and economic instruments and other measures, promoted by the Member States, in order to provide incentives for the application of the waste hierarchy, such as landfill and incineration charges, pay-as-you-throw schemes, extended producer responsibility schemes, and incentives for local authorities [50–54], or other appropriate instruments and measures[26];
- measures, such as taxation of virgin materials[27] [56];
- oriented and aggregated demand for secondary materials [57];
- measures, promoted by the Member States, taking into account the impact of products throughout their life cycle and aiming to encourage the development, production and marketing of products that are suitable for multiple use [58–60],

national or regional schemes it is important to secure that: there is no conflict with the European harmonised approach; no technical barriers to trade are invoked; impacts on costs and administrative burden have been fully taken into account and where possible mitigated; innovative companies are not put at a disadvantage compared to other companies".

[23]Advanced sorting experiments are described in Linß et al. [26]: "*Sensor-based single particle sorting devices are advanced techniques to sort efficiently usable material fractions and to discharge impurities and contaminants from the recyclable fractions. (...) The particles are transported by a conveying system and pass a push-broom based camera system which measures the spectra in reflection mode. The particles are then sorted by compressed air pulses*".

[24]The Hyperspectral-Imaging (HSI) technique for the selectively recognition of contaminated particles for an industrial implementation for automatic sorting and separation lines for C&DW recycling is described in Hollstein et al. [45].

An experimentation for coupling Data from hyperspectral cameras and reflectance sensors and the robotic technology for an innovative CDW ro sorting system is described in Zerbi1 and Landò1 (2017).

[25]In Flanders, a new quality management and traceability system has been developed by Flemish construction confederation. This monitoring system aims at the certification of waste streams that originate from selective demolition. "*The selectivity of the demolition is driven by the economic incentives of (high landfill fees and differential gate fees at recycling plants) and obligations from Flemish/Belgian legislation (e.g. mandatory elimination of asbestos and other hazardous materials). These incentives make "semi-selective demolition" current practice in the Flemish region. We define semi-selective demolition as a demolition work where the demolition company selectively collects all hazardous substances and that part of the non-hazardous substances that would overly reduce the quality of the stony fraction. The selective collection of the latter is determined by their value, the acceptance policy of the crushing installations of the stony fraction and by the time consumed for selective removal*" [44].

[26]See directive (EU) 2018/851 [55].

[27]Experience in the UK shows that heavier taxation of virgin materials improves recycling levels for construction and demolition waste so that virgin resources are saved [5].

that contain recycled materials, that are technically durable and easily reparable and that are, after having become waste, suitable for preparing for re-use and, where appropriate, potential for multiple recycling[28];

- available and reliable data regarding the quality and quantity of CDW;
- available and reliable quality assurance methods, procedures, testing technics,[29] voluntary schemes[30] and standards[31] regarding the properties of recycled materials;
- a critical mass of competing companies, specialized in waste management, with high skills and provided with advanced equipment;
- economic instruments and other measures, promoted by the Member States, in order to provide incentives for the application of the waste hierarchy, such as landfill and incineration charges, pay-as-you-throw schemes, extended producer responsibility schemes, facilitation of food donation, and incentives for local authorities, or other appropriate instruments and measures;

[28]See directive (EU) 2018/851 [55].

[29]To give an example, the image analysis technique for the estimation of the water content and the fresh properties of concretes made with Recycled Aggregates Concretes is described in Juez and Artoni [61]. See also Chiranjib Roy et al. [62].

[30]To give some examples:

- the European Commission's Joint Research Centre (JRC), is currently developing the document "Best Environmental Management Practice in the Waste Management Sector", which will cover three waste streams: C&D waste, municipal solid waste and medical waste. The document will cover the following waste activities: waste management, prevention, re-use, collection and treatment. See: The Joint Research Center, 2016, http://susproc.jrc.ec.europa.eu/activities/emas/index.html;
- the voluntary French Management and Audit Scheme, QUALIRECYCLE BTP, is a management scheme developed by Syndicat des Recycleurs du BTP (SR BTP) for waste management companies in order to evaluate, report and improve their performance in the compliance, environment and safety fields and show their commitment to recovering issues. See: SR BTP, www.recycleurs-du-btp.fr/quali-recycle-btp/inFrench;
- the European Panel Federation (EPF) has developed an industry standard that describes the conditions under which recycled wood can be accepted for the manufacturing of wood-based panels. This standard comprises general requirements on quality and chemical contamination, classes of unacceptable materials (e.g. wood treated with PCP) as well as reference methods for sampling and testing. See: European Panel Federation (EPF) 2016, www.europanels.org.

[31]See the specific Belgian standardisation and certification framework, based on the European standard EN 206. "*Specific aspects of the Belgian approach are the implementation of the 'specificity of use'-principle, application of correction and safety factors on the calculation of the effective water-to-cement-ratio and the required quality of the recycled aggregates. To support the development of the standard, research is being executed in several Belgian laboratories*" [38].

- data-driven solutions, which are now spreading in many States, able to improve the cost-effectiveness CDW management on site,[32] such as real-time monitoring of CDW generated on the construction site and automatic compliance checks.

References

1. SEC (2011) 1067, Commission staff working paper analysis associated with the roadmap to a resource efficient Europe. Part II Accompanying the document communication from the commission to the European Parliament, the council, the European economic and social committee and the committee of regions Roadmap to a Resource Efficient Europe
2. European Construction Sector Observatory (2019) EU construction sector: in transition towards a circular economy, Trend Paper Series, April 2019
3. Augiseau V, Barles S (2017) Studying construction materials flows and stock: a review. Resour Conserv Recycl 123:153–164
4. Iacoboaea C, Aldea M, Petrescu F (2019) Construction and demolition waste-a challenge for the European Union? Theor Empir Res Urban Manag 14(1):30–52
5. Stephan A, Athanassiadis A (2017) Quantifying and mapping embodied environmental requirements of urban building stocks. Build Environ 114:187–202
6. Stephan A, Athanassiadis A (2018) Towards a more circular construction sector: Estimating and spatialising current and future non-structural material replacement flows to maintain urban building stocks. Resour Conserv Recycl 129:248–262
7. Condeixa K, Haddad AN, Boer D (2017) Material Flow Analysis of the residential building stock at the city of Rio de Janeiro. J Clean Prod 149:1249–1267
8. Gontia P, Nageli C, Rosado L, Kalmykova Y, Osterbring M (2018) Material-intensity database of residential buildings: a case-study of Sweden in the international context. Res, Conserv Recycl 130:228–239
9. Kalmykova Y, Sadagopan M, Rosado L (2018) Circular economy—from review of the theories and practices to development of implementations tools. Res, Conserv Recycl 135:190–201
10. Pomponi F, Moncaster A (2016) Circular economy for the built environment: a research framework. J Clean Prod 143:710–718

[32]As an example [37], BRE SMARTWaste is a flexible, online-reporting platform for companies of all sectors that can support in managing and reducing waste outputs, impacts and costs. It is intended for clients, contractors, owners, operators and occupiers. It can be used to prepare, implement and monitor site waste management plans (SWMP), that describe how materials will be managed efficiently and disposed of legally during construction, explaining how the re-use and recycling of materials will be maximized. This involves estimating how much of each type of waste is likely to be produced and the proportion of this that will be re-used or recycled on site, or removed from the site for re-use, recycling, recovery or disposal.
SMARTWaste can support to:

- Save time tracking and reporting against company sustainability targets.
- Compare regional and project performance.
- Give access to all project information in one place.
- Reduce waste, energy and water costs.
- Set and monitor targets to reduce site based impacts.
- Obtain BREEAM and Code for Sustainable Homes credits.

See: https://www.designingbuildings.co.uk/wiki/BRE_SMARTWaste_online_reporting_platform.

11. EU (2018) Guidelines for the waste audits before demolition and renovation works of buildings. EU construction and demolition waste management
12. COM (2018) 656 Report from the commission to the european parliament, the council, the european economic and social committee and the committee of the regions on the implementation of EU waste legislation, including the early warning report for Member States at risk of missing the 2020 preparation for re-use/recycling target on municipal waste
13. EC (2017) Resource efficient use of mixed wastes. Improving management of construction and demolition waste. Final report
14. EEA (2012a) Report No 7/2012 Movements of waste across the EU's internal and external borders, Report
15. EC (2000) Commission decision of 3 May 2000 replacing decision 94/3/EC establishing a list of wastes pursuant to Article 1(a) of Council Directive 75/442/EEC on waste and council decision 94/904/EC establishing a list of hazardous waste pursuant to Article 1(4) of Council Directive 91/689/EEC on hazardous waste, European Commission (OJ L 226), 6 September 2000
16. ETC/SCP (2012) Transboundary shipments of waste in the European Union—reflections on data, environmental impacts and drivers. Prepared by Fischer C, Junker H, Paleari S, Zoboli R, Wuttke J (eds) ETC/SCP working paper 2/2012, European topic centre on sustainable consumption and production
17. DYNAMIX (2013) The underlying reasons for resource (in)efficiencies. Exec Summ Deliv 2:2
18. EEA (2018) Report No 4/2018, Waste prevention in Europe policies, status and trends in reuse in 2017
19. Bauteilbörse Bremen (2018) (http://bauteilboersebremen.de)
20. BS EN ISO 14044:2006 Environmental management—Life cycle assessment—Requirements and guidelines
21. Andrade JB, Bragança L (2019) Assessing buildings' adaptability at early de-sign stages. In: IOP Conference Series: Earth and Environmental Science, vol 225, No 1, p 012012). IOP Publishing
22. Talamo C (2013) "Design for Maintainability. tools and procedures, in changing needs, adaptive buildings, smart cities, adaptive buildings, smart cities. In: Proceedings of the 39th congress of IAHS "The International Association for Housing Science" Changing Needs, Adaptive Buildings, Smart Cities, September 2013, Milano, pp 463–472
23. Eray E, Sanchez B, Haas C (2019) Usage of interface management system in adaptive reuse of buildings. Buildings 9(5):105
24. Gicala M, Sagan J, Sobotka A (2019) Decision support in the process of choosing a residential building technology. In MATEC Web of Conferences, vol 262, p 07003. EDP Sciences
25. Navarro-Rubio J, Pineda P, Garcia-Martinez A (2019) Sustainability, prefabrication and building optimization under different durability and re-using scenarios: potential of dry precast structural connections. Sustain Cities Soc 44:614–628
26. Martani C, Paganin G, Talamo C (2013) The control, at the design stage, of risks related to buildings management over time, TECHNE 6, 2013. ISSN 109–116:2239–2243
27. Talamo C, Lavagna M (2010) The role of durability and planned maintenance in the environmental assessment of buildings, Design, technology, refurbishment and management of buildings. In: Proceedings of the 37th IAHS world congress on housing science, Santander (Spain), pp 1–8
28. Akanbi L, Oyedele L, Davila Delgado JM, Bilal M, Akinade O, Ajayi A, Mo-hammed-Yakub N (2019) Reusability analytics tool for end-of-life assessment of build-ing materials in a circular economy. World J Sci Technol Sustain Dev 16(1):40–55
29. Rasmussen FN, Birkved M, Birgisdottir H (2019, January). Upcycling and design for disassembly-LCA of buildings employing circular design strategies. In Iop conference series: earth and environmental science, vol 225, No 1, p 012040. IOP Publishing
30. WRAP (2010) Designing out waste tool for buildings reference guide (WRAP Project PRO 084-001)

31. BS 8887: 2009, Design for manufacture, assembly, disassembly and end-of-life processing (MADE) Part 2: Terms and definitions
32. EEA (2017) Report No 6/2017, Circular by design Products in the circular economy
33. Hobbs G, Adams K (2017) Reuse of building products and materials–barriers and opportunities. In: International HISER conference on advances in recycling and management of construction and demolition Waste 21–23 June 2017, Delft University of Technology, Delft, The Netherlands
34. Kretsloppsrådet (2015) Resource and waste guidelines during construction and demolition, November 2015
35. EC (2016a) Technical and economic study with regard to the development of specific tools and/or guidelines for assessment of construction and demolition waste streams prior to demolition or renovation of buildings and infrastructures. Final report of EU Specific Contract 30-CE-0751644/00-00–SI2.720069 (2016)
36. EC (2018) Guidelines for the waste audits before demolition and renovation works of buildings
37. EC (2019) European construction sector observatory, EU construction sector: in transition towards a circular economy, Trend Paper Series
38. Wahlström M, Castell-Rüdenhausen MZ, Hradil P, Hauge-Smith K, Oberender A, Ahlm M, Hansen JB (2019) Improving quality of construction and demolition waste-requirements for pre-demolition audit
39. EEA (2015) Waste prevention in Europe—the status in 2014, Report No 6/2015
40. Talamo C, Bonanomi M (2015) Knowledge management and information tools for building maintenance and facility management, Springer
41. Linß E, Karrasch A, Landmann M (2017) Sorting of mineral construction and demolition wastes by near-infrared technology. In: International HISER conference on advances in recycling and management of construction and demolition waste 21–23 June 2017, Delft University of Technology, Delft, The Netherlands
42. Rose C, Stegemann J (2018) From waste management to component management in the construction industry. Sustainability 10(1):229
43. RWS (2015) Circular economy in the Dutch construction sector, report, 18 December 2015
44. Bergmans J, Dierckx P, De Regel S, Eynde AV, Van de Velde P, Broos K (2017) Semi-selective demolition: current demolition practices in flanders. In: International HISER Conference on Advances in Recycling and Management of Construction and Demolition Waste 21–23 June 2017, Delft University of Technology, Delft, The Netherlands
45. Hollstein F, Wohllebe M, Herling M, Cacho I, Arnaiz S (2017) Sorting of construction and demolition waste by hyperspectral-imaging. In: International HISER conference on advances in recycling and management of construction and demolition waste 21–23 June 2017, Delft University of Technology, Delft, The Netherlands
46. Commission Decision 2000/532/CE of 3 May 2000 replacing decision 94/3/EC establishing a list of wastes pursuant to Article 1(a) of council directive 75/442/EEC on waste and council decision 94/904/EC establishing a list of hazardous waste pursuant to Article 1(4) of Council Directive 91/689/EEC on hazardous waste
47. Vrijders J, Dooms B, Van Itterbeeck P (2017) Putting recycled concrete into practice in Belgium: from case studies over chain management to standards and quality assurance. In: International HISER conference on advances in recycling and management of construction and demolition waste 21–23 June 2017, Delft University of Technology, Delft, The Netherlands
48. Williams TG, Heidrich O, Sallis PJ (2010) A case study of the open-loop recycling of mixed plastic waste for use in a sports-field drainage system. Resour Conserv Recycl 55(2):118–128
49. EC (2016b) EU Construction and demolition waste management protocol, September 2016
50. Ellen MacArthur Foundation, the McKinsey Centre for Business and Environment and the Stiftungsfonds für Umweltökonomie und Nachhaltigkeit (SUN), Growth within: a circular economy vision for a competitive Europe, report 2015
51. European Commission (2012) directorate-general environments, guidance on the interpretation of key provisions of directive 2008/98/EC on waste
52. European Commission (2010) Being wise with waste: the eu's approach to waste management
53. European Innovation Partnership (EI) (2016) Raw materials, raw materials scoreboard

54. European Parliament Resolution of 9 July 2015 on resource efficiency: moving towards a circular economy (2014/2208(INI)
55. Directive (EU) 2018/851 of the European Parliament and of the council of 30 May 2018 amending directive 2008/98/EC on waste
56. EEA (2012) European innovation partnership on raw materials, raw materials scoreboard, European Union, 2016
57. SWD (2019) 90 commission staff working document accompanying the document report from the commission to the European Parliament, the council, the European economic and social committee and the committee of the regions on the implementation of the Circular Economy Action Plan
58. Decision No 1386/2013/EU of the European Parliament and of the council of 20 November 2013 on a general union environment action programme to 2020 'Living well, within the limits of our planet'
59. Directive 2009/125/EC of the European Parliament and of the council of 21 October 2009 establishing a framework for the setting of ecodesign requirements for energy-related products
60. EEA (2016) Circular economy in Europe. Developing the knowledge base, Report
61. Juez JM, Artoni R, Cazacliu B (2017) Monitoring of concrete mixing evolution using image analysis. Powder Technol 305:477–487
62. Chiranjib Roy, Md. Shariful Islam, Firoz Ahamed, Shape parameters of recycled aggregate and determination its area of mortar and aggregate using imaging techniques. Int J Adv Eng Technol 3(1):31–35 (January 2019; ISSN: 2456-7655)
63. Decision No 1600/2002/EC of the European Parliament and of the council of 22 July 2002 laying down the sixth community environment action programme
64. EEA (2019) Report No 02/2019, Preventing plastic waste in Europe

Websites

65. https://eea.europa.eu/data-and-maps/figures/trade-in-selected-waste-materials
66. http://appsso.eurostat.ec.europa.eu/nui/show.do?dataset=env_ac_rmefd&lang=en
67. http://scp.eionet.europa.eu/
68. http://wrap.org.uk/nwtool
69. http://frnorg.uk
70. http://btbbasel.ch/ueber-uns.html
71. http://cdrecycler.com/article/frost-sullivan-report-construction-demolition-market/
72. https://umweltbundesamt.de/sites/default/files/medien/461/publikationen/texte_56_2010_kurzfassung_e1_0.pdf
73. http://europanels.org
74. https://designingbuildings.co.uk/wiki/BRE_SMARTWaste_online_reporting_platform

Chapter 3
Reuse as a Bridge Between Waste Prevention and the Circular Economy

Abstract The chapter deals with the subject of reuse, considered in the perspective of waste prevention. The chapter analyses the Member States approaches towards reuse and highlights the barriers and drivers that may influence its development as a widespread practice. Besides, the reuse businesses is considered also in the perspective of a range of "low-impacts" processes [1], that we could we could call "re"-processes. These processes, all starting from deconstruction/disassembly activities, can be identified as: rebuild, reclaim, recondition, reconstruct, refurbish, remanufacture, renew/renovate, repair, repurpose, restore, rework. In particular the chapter focuses of the remanufacturing practices.

Keywords Reuse · Remanufacturing · Information

3.1 Waste Prevention and Reuse

Notwithstanding the many barriers, described in the previous chapter (see Chap. 2), circular economy is surely boosting a more efficient use of the resources, opening new markets and creating new skills and business, considered from the perspective both of prevention of waste and of generation of secondary products.

Considering Construction and Demolition Waste (CDW), the subject of the waste hierarchy (Dir. 98/2008) (see Chap. 1), that is the priority order in waste prevention and management, should deserve some further reflections, focusing on the different impacts related to reuse and recycling.

EU legislation does not identify specific targets for reuse and recycling, and much of waste streams, diverted from landfill, has been recycling, generating mainly lower value products (downcycling) [2].

According to the EEA report "*More from less. Material resource efficiency in Europe: 2015 overview of policies, instruments and targets in 32 countries*" [3], the

This chapter is authored by Cinzia Talamo, Politecnico di Milano, Department of Architecture, Built Environment and Construction Engineering, Milan, Italy. e-mail: cinzia.talamo@polimi.it.

M. Migliore et al., *Strategies for Circular Economy and Cross-sectoral Exchanges for Sustainable Building Products*, Springer Tracts in Civil Engineering,
https://doi.org/10.1007/978-3-030-30318-1_3

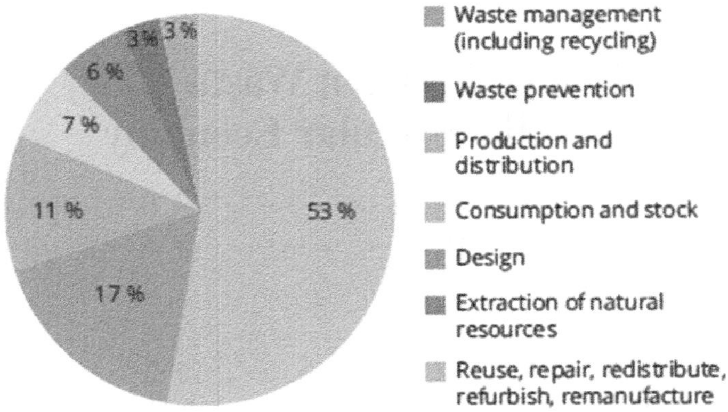

Fig. 3.1 National policy approaches to closing material loops according to the study "*More from less—material resource efficiency in Europe*" of European Environment Agency. *Source* EEA [3]

circular economy approaches in the Member States (MS) appear to be still predominantly focused on the waste stage, rather than on the early stages of the product life cycle. According to the EEA report [3], investigating all sectors, only a few countries consider reuse, repair, redistribution, refurbishment, remanufacturing or design as main instruments for closing material loops (Fig. 3.1).

As Rose and Stegemann [4] argue, there is a danger in assuming that the impact of CDW has been successfully mitigated as recycling rates rise. "*A first problem is that these data are based on whether waste is sent to recycling companies, rather than whether it is recycled in reality. Secondly, the impacts of transportation and recycling processes can be considerable. Thirdly, recycling processes can be highly wasteful*" [4].

Hobbs and Adams [5] underline that reuse should be considered as a priority, if compared to recycling, because reuse typically requires minimal processing before reapplication in a similar application, whereas recycling typically requires breaking down waste into a homogeneous material for a lesser value application or introduction as replacement feedstock for manufactured components. "*A common misunderstanding lies between the realms of reuse and recycling of old buildings; they are often considered together when they are actually competing choices for the continuing use of resources*" [5].

Starting from this assumption, Hobbs and Adams [5] highlight the many benefits of reuse: according to the Waste Framework Directive, reuse can effectively remove materials and products from the waste stream and allow to use materials and products for further applications without the regulatory restrictions, that can accompany recycling. Certification is not always required to enable reuse and the heritage and environmental value of the products still remain in circulation.

Bougrain and Laurenceau [6] propose a further point of view regarding the possible environmental and social benefits of reuse. The activities of deconstruction and disassembly,[1] related to reuse, imply surely less environmental impacts then demolition and, being often highly labor intensive, can represent opportunities for reintegrated employees and artisanal small businesses as well as for the economic revitalization of less-developed areas.

The EEA report "*Waste prevention in Europe*" [1] effectively highlights the potential environmental and socio-economic benefits of reuse by describing it as a bridge between waste prevention and the circular economy, arguing that: "*reusing products and components at the end of their use phase can reduce waste generation and potentially save natural resources by extending the use phase of products at the same time. Reuse, as well as preparation for reuse, can thus provide a link between the waste hierarchy of the Waste Framework Directive, on the one hand, and the European Commission's Circular Economy Action Plan, on the other*".

Nevertheless, the European Union's approach to waste recognizes to reuse a high value. The Waste Framework Directive places preparing for reuse at the second level of the waste hierarchy and calls on the Member States (article 11) to take appropriate measures to promote the reuse of products and the preparing for reuse activities, notably by encouraging and boosting supports such as reuse and repair networks, economic instruments, procurement criteria, etc.

Besides, the Waste Framework Directive required Member States to establish waste prevention programmes. The Directive provides flexibility regarding the nature of the programmes, but it requires that objectives and qualitative or quantitative benchmarks are set. Other frameworks, related to reuse are the Roadmap to a Resource Efficient Europe and the EU's Seventh Environment Action Programme, which also recognise the need for waste prevention.

A new framework for waste policy and resource efficiency was introduced in the Circular Economy Action Plan COM (2015) 614 [7], (see Chap. 1). This action plan addresses a range of economic sectors, including waste and deals with the whole cycle: from production and consumption to waste management and the market for secondary raw materials[2] [8]. Within the scope of the Circular Economy Action Plan, the Commission announced that it "*will promote the reparability, upgradability, durability, and recyclability of products by developing product requirements relevant to the circular economy in its future work under the Ecodesign Directive*" [7].

Later, the Directive (EU) 2018/851 [9], emending the Directive 2008/98/EC [10] on waste (see Chap. 1), includes, re-use between the measures for the prevention of waste: "*encourage the re-use of products and the setting up of systems promoting repair and re-use activities, including in particular for electrical and electronic equipment, textiles and furniture, as well as packaging and construction materials and products*" (Dir. 2018/851).

---

[1] According to Bougrain and Laurenceau [6] disassembly means: "*taking apart components without damaging, but not necessarily to reuse them*" while "*deconstruction is similar to disassembly but with thoughts towards reusing the components*". Conversely, the action of demolition is "a process of intentional destruction" that generates waste that can be recycled.

[2] See also (http://ec.europa.eu/environment/circulareconomy/index_en.htm) for circular economy.

The Directive also states that by December 2024, the Commission shall consider the setting of preparing for reuse and recycling targets for construction and demolition waste and its material-specific fractions, textile waste, commercial waste, non-hazardous industrial waste and other waste streams, as well as preparing for reuse targets for municipal waste and recycling targets for municipal bio-waste. To that end, the Commission shall submit a report to the European Parliament and to the Council, accompanied, if appropriate, by a legislative proposal.

The EEA report [1] is focused on reuse and related waste prevention activities (Table 3.1). It investigates 33 national and regional waste prevention programmes, adopted by the end of 2017, and gives an overview of the European approaches to reuse practices. The report denounces also that national approaches towards reuse are very diverse and rely mostly on voluntary arrangements:

- all the 33 waste prevention programmes, and related measures, analysed by the report, describe a scenario of a broad range of products as well as stakeholders;
- many countries have initiated reuse networks that offer high-quality second-hand products to consumers (e.g. Flanders and Austria);
- 47% of countries adopts voluntary agreements;
- 35% are developing informational instruments specifically related to the support of reuse and reuse systems;
- 10% of the programmes implement or aim to implement regulatory measures;
- only 8% use or plan to use economic instruments.

The development of the practice of reuse depends also on the level of maturation of various aspects, such as: the modalities the products exchange/trade and their supporting infrastructures, the collection methods and the models of governance [1], for example:

- citizen to citizen (e.g. online marketplaces, etc.) (C2C);
- citizen to charity organisations, reuse businesses (e.g. with drop-off centres for used goods) or (separate) municipal waste collection [e.g. municipal 'dumps' or kerbside collection of waste electrical and electronic equipment (WEEE)] ('C2B');
- social or other enterprises to people with demand for certain reuse products ('B2C') (e.g. furniture);
- social enterprises or other reuse businesses to enterprises with demand for reuse products ('B2B') (e.g. building components exchange);
- both public and private reuse centres, as requested by the Waste Framework Directive[3];
- platforms for the matching of demand and offer.

In Table 3.2 a summary of the possible actions to support reuse and some related cases of application in some Member States (MS).

---

[3]Between the examples of waste prevention measures referred to the article 29, the Waste Framework Directive includes: "*the promotion of the reuse and/or repair of appropriate dis-carded products or of their components, notably through the use of educational, economic, logistic or other measures such as support to or establishment of accredited repair and reuse centres and networks especially in densely populated regions*".

Table 3.1 Objectives linked to reuse included in national and regional waste prevention programmes

Nations and regions	Objectives and scope	Reuse indicators
Austria	Further expansion and consolidation of the reuse networks, expansion of the reuse collection of old electrical appliances in communities, expansion of the reuse collection to other usable goods and creation of further markets for reuse products	Reuse: collected quantity of reusable products per year, number and turnover of reuse companies, number of second-hand products sold annually, rate of reuse of construction waste'
Belgium/flanders	Five objectives for sustainable materials management in the building sector, one of them requiring design to enable the reuse of materials	Directly referring to quantitative targets
Croatia	Establish reuse centres and encourage exchange and reuse of used products, and increase the reuse of demolition material	The number of newly opened workplaces dealing with waste reuse
Czech republic	Support reuse and service centres and charitable organisations for the repair and reuse of products and materials	The amount of collected textiles, footwear and selected reusable products, t/year' – Number of service centers and network for life extension and reuse of products and components. Number of products that have gone through service centers and were reused' – The number of non-profit organisations developing activities for re-use of products and activities related to waste prevention
Denmark	Facilitate the reuse and recycling of electronics and electronic waste to extend product lifespans	Consumption of clothes and textiles, including second-hand clothes'
Estonia	Increase customer awareness of the potential contribution they can make to waste prevention	Quantity of waste prepared for reuse'

(continued)

Table 3.1 (continued)

Nations and regions	Objectives and scope	Reuse indicators
Finland	Increase reuse through public procurement, town planning, education and advice, promoting the reuse of building supplies and components, researching potential economic policy tools to prolong the lifespan of products, studying the potential for and hindrances to reusing packaging, increasing the reuse of electrical and electronic equipment (EEE) and boosting the supervision of illegal WEEE shipments	Eee reuse volume, tonne/year
Germany	Support reuse of products	– Share of reused electronic products' – Share of reusable packaging'
Greece	Promote reuse of electrical and electronic equipment.	– Number of reuse centres' – Share of reusable packaging'
Ireland	Support sustainable growth and employment in the green economy, including reuse enterprises	Not specifically referring to reuse
Italy	Not specifically referring to reuse	Number of products that enter and leave a reuse centre, and number of visits made to reuse centres
Luxembourg	Prevent waste and guide consumers towards products with greater longevity or multiple uses	None
Malta	Promote reuse and repair initiatives	None
Poland	Increase reuse, for example through networks for the exchange and repair of electrical and electronic equipment and by collecting and preparing WEEE for reuse	– 'Percentage of packaging placed on the market that is reusable' – 'Percentage of the total mass of waste equipment collected in a given year that is totally reused'
Portugal	Reduce the amount of waste through increased material resource efficiency and reuse	Not specifically referring to reuse
Slovenia	Promote reuse of items, materials or products	Not specifically referring to reuse
Spain	Reuse of furniture, clothing and textiles as well as electronic equipment	Number of operative reuse centres and number of associated new jobs'

(continued)

Table 3.1 (continued)

Nations and regions	Objectives and scope	Reuse indicators
UK-England	Encourage better valuing of resources through awareness-raising among businesses and the public regarding waste reduction options, using products for longer, repairing broken items and enabling the reuse of items by others	Not specifically referring to reuse
UK-Northern Ireland	Increase the supply of and demand for good-quality reusable items by improving collection, promotion and public procurement	Not specifically referring to reuse
UK-Scotland	Expand the reuse sector: 'we want the sale and use of second hand goods to be seen as an attractive, mainstream, good value option for an increasing range of products. We want reuse businesses and community organisations to thrive, on the back of a growing reputation for quality and value for money'	Not specifically referring to reuse
UK-Wales	Support for households and businesses to reduce their quantity of waste through reuse	Not specifically referring to reuse

Source EEA [1]

3.2 The "Re"-processes

When dealing with the reuse businesses, we should consider various types of second hand building products, that can be obtained through a range of "low-impacts" processes [1], that we could we could call "re"-processes. These processes, all starting from deconstruction/disassembly activities [6], can be identified as: rebuild, reclaim, recondition, reconstruct, refurbish, remanufacture, renew/renovate, repair, repurpose, restore, rework.

It is important to state a precise and shared definition for each of these processes, as "*the lack of definition can create barriers to international trade, as products and core (a used part intended to become a remanufactured product) are sometimes considered as waste, rather than potentially high value inputs into a (re-)manufacturing process. Consumer perception of remanufactured products could also be affected by this lack of definition*" [12].

Table 3.3 reports a list of "re"-processes and related definitions. Many of these

Table 3.2 Actions boosting reuse

Reuse action	Rationale
Mandatory pre-demolition and renovation audits with promotion of reuse	There are requirements to undertake pre-demolition audits in a number of MS, such as Hungary and Finland. However, carrying out such an audit does not necessarily result in increased reuse, especially where the rationale behind it is for identifying hazardous substances. If a common approach was adopted, whereby a third party conducted the audit and certain levels of reuse were identified and measured against, this might have a positive effect. Subsequent linking with recertification and a suitable market demand would add weight to such audits
Managing supply and demand	Frequently the specifications of the follow-on development does not enable reusable demolition products to be incorporated at the same site. Several countries have waste exchanges and industrial symbiosis programmes to help match supply and demand—although these platforms are equally used to facilitate recycling. The ability to match supply and demand for reusable products and materials is essential, as evidenced by reclamation activities in the UK through the Salvo network. There is a clear need to have traceability in the supply chain, linked to the information requirements of the markets to which they can be best used
Innovation in reuse	Removing products for reuse during demolition can be time consuming and increase health and safety risks, compared to mechanical techniques to separate into material streams. There are innovations, such as REBRICK in Denmark that can help overcome such barriers. The EU R&D programme could have a focussed call to develop additional technologies targeting reuse on a pan EU collaborative basis. Transfer of existing technologies could also be promoted, taking into account the typically SME nature of the reclamation industry
Support for the reclamation sector	In addition to supporting R&D in this sector, there could be advantages in supporting those in this sector, both existing and to establish new facilities. These sites act as stockholders for products and materials, enabling their accumulation and retention for a demand that might not be available when they are removed from buildings at end of life. Making public land available for such enterprises at a reasonable cost could help new business start-ups/social enterprises. Existing facilities could also be supported, upon the condition that they had minimal 'reproduction' stock
Construction Product declaration and recertification	Lack of certification and uncertainty over performance prevent reclaimed products and materials being used in mainstream construction, where there could be significant market pull. The key issue, beyond availability of certain volumes, relates to demonstration

(continued)

Table 3.2 (continued)

Reuse action	Rationale
Better impact data	There is no clear impact data that can be used to promote reuse in preference to recycling. These are typically a combined option on LCA. This is despite the widespread opinion that reuse offers better environmental, social and (possibly) economic outcomes. Such evidence, or calculation tool, could be valuable in green procurement. EU R&D projects focussed on existing buildings and optimised demolition to promote reuse have impact data aspects, including HISER and FISSAC
Data management, including BIM	The Austrian Building Pass is a good example of where the importance of transferring information across a building life cycle has been recognised by a MS. This approach could be adopted across the EU with far reaching positive effects. A vehicle that could be sensibly used to enable such a pan EU policy would be through the combining of resource optimisation data within building information modelling. This is a key output of the EU funded R&D project—Buildings as Materials Banks (BAMB) Building Information Management (BIM) is a key way of storing information on a building digitally, such as the design parameters and type and amount of products and materials as well as management data. It is being increasingly used to manage a building's data across the lifecycle. Two EU funded R&D projects are developing BIM-enabled tools to take into account resource management data; these are: Buildings as Materials Banks (BAMB) and the H2020 HISER project (Holistic Innovative Solutions for an Efficient Recycling and Recovery of Valuable Raw Materials from Complex Construction and Demolition Waste)

Source EC [11]

processes share the aim to extend the lifespan of used products and core by facing the phenomenon of degradation[4] and/or obsolescence.[5]

The outcome of these processes is a product (Fig. 3.2):

- whose performances may be lower, equal or better than the ones of the original condition;
- that can be used in the same site by the original user or sold/exchanged in a second hand marketplace;
- that can maintain the same original function or be configured for a diverse use with various forms of warranty.

Between the "re"-processes, the process of remanufacturing can be considered the most complex and the one able to deliver goods, that are equal or better than

[4]Degradation: Detrimental change in physical condition, with time, use or external cause. Note 1 Degradation may lead to a failure. Note 2 In a system context, degradation may also be caused by failures within the system [13].

[5]Obsolescence: State of an object, determined by a process of transformation, that causes a situation of non-compliance between requirements and provided performances.

Table 3.3 "Re"-processes definitions

"Re"-processes	Definitions
Rebuild	To repair by taking apart and reconstructing, often with new parts to a standard as close as possible to its original condition regarding appearance, performance and life expectancy Typically used for car chassis and historical industrial installations, without major engineering changes [14]
Recondition	To return a used product to a satisfactory working condition by rebuilding or repairing major components that are close to failure, even where there are no reported or apparent faults in those components With respect to reconditioning: – Manufacturing effort involves the replacement of worn or broken parts, generally less extensive than required to remanufacture, but more than necessary for repair – Performance after reconditioning is expected to perform its intended role but the overall performance is likely to be inferior to that of the original model – Any subsequent warranty is generally less than new or a remanufactured product but the warranty is likely to cover the whole product (unlike repair); reconditioned products do not require a warranty equivalent to that of a newly manufactured equivalent [15] A process of returning a product to good working condition by replacing major components that are faulty or close to failure, and making 'cosmetic' changes to update the appearance of a product, such as cleaning, changing fabric, painting or refinishing. Any subsequent warranty is generally less than that issued for a new or a remanufactured product, but the warranty is likely to cover the whole product (unlike repair). Accordingly, the performance of the product may be less than as–new [16] The potential adjustment to components bringing an item back to working order, though not necessarily to an 'as new' state [17]
Reclaim	Recovery from waste (useful substances), recovery of materials [14]
Reconstruct	Restoring a product into its original or complete condition by re-building/re-constructing/re-assembling it from remaining parts. Defects can be corrected, and the structure may change [14]
Refurbish	Aesthetic maintenance of a product to bring it back to a condition comparable to that of a new product. It does not include repairing or rebuilding any components Use for polishing of historical inventory, changing the optical appearance of used parts (e.g. consumer electronic) [14] The largely aesthetic improvement of a product which may involve making it look like new, with limited improvements to functionality [17]

(continued)

Table 3.3 (continued)

"Re"-processes	Definitions
Remanufacture	To return a used product to at least its original performance with a warranty that is equivalent or better than that of the newly manufactured product *NOTE 1 From a customer viewpoint, the remanufactured product can be considered to be the same as the new product* *NOTE 2 With respect to remanufacture*: – *Manufacturing effort involves dismantling the product, the restoration and replacement of components and testing of the individual parts and whole product to ensure that it is within its original design specifications* – *Performance after remanufacture is expected to be at least to the original performance specification* – *Any subsequent warranty is generally at least equal to that of new product* *NOTE 3 This assumes that remanufacture applies to like-for-like products* [15] A remanufactured product fulfils a similar function to the original part. It is manufactured using a standardized industrial process in line with specific technical specifications. The industrialized process incorporates defined core management standards. A remanufactured spare part is warranted as a new spare part [18]
Renovate/renew	Lat. renovare "to make new". Damage due to wear from ordinary use is removed; item is restored to its original condition Renovation is mostly used to describe the action of repairing a building [14] Renew: To make new or as if new again[a]
Repair	Returning a faulty or broken product or component back to a usable state A repair may use remanufactured or reconditioned parts With respect to repair: – manufacturing effort is the minimum required to address the specified fault – after repair, the product is expected to be in a useable state, but assurances of performance are generally limited to the repaired part – any subsequent warranty is generally less than that of newly manufactured, remanufactured or reconditioned equivalents and may apply only to the component that has been replaced or repaired [15] Unlike rework, repair can affect or change parts of the nonconforming product [19] Fixing a fault but with no guarantee on the product as a whole [17]
Repurpose	To use a product for a different purpose than originally intended NOTE 1 An item can be repurposed by modifying it to fit a new use, or by using the item as it is in a new way[b] To utilize a product or its components in a role that it was not originally designed to perform NOTE 1 This action deals specifically with products and assemblies and not materials, which falls under recycling NOTE 2 Augmentation of the product may be required to fulfil its new role [15]

(continued)

Table 3.3 (continued)

"Re"-processes	Definitions
Restore	Restoration includes measures to preserve the material existence and the physical characteristics of an object NOTE: Restoration is mostly used in the context of works of art or of general objects of cultural history [14]
Reuse	Operation by which a product or its components are put back into use for the same purpose at end-of-life [15] The simple reuse of a product with no modifications. Operation by which a product or its components are put back into use for the same purpose at end of life [17]
Rework	Action on a nonconforming product to make it conform to the requirements. Rework includes all follow-on efforts such as disassembly, repair, replacement, reassembly, etc. NOTE Rework is a generic term for repair. In contrast to repair, rework can neither affect nor change parts of the nonconforming product. Rework can be done also restarting the production process from row material level [19]

[a]www.thefreedictionary.com/renew
[b]http://www.investopedia.com/terms/r/repurposing.asp#axzz1g2lpWQPR, 09.12.2011

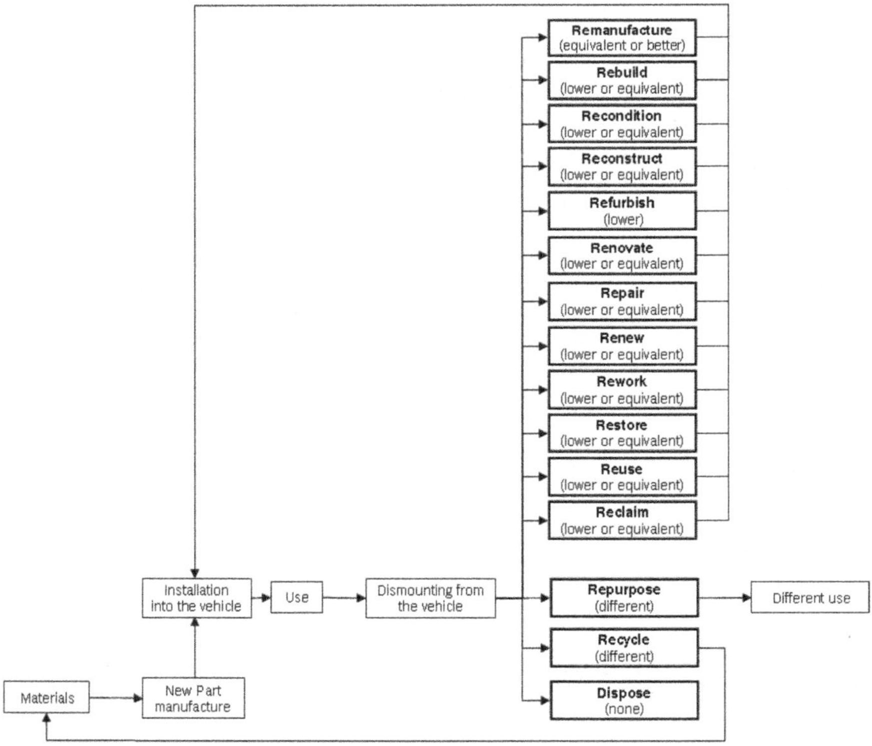

Fig. 3.2 The "re"-processes. *Source* APRA [14]

the original products, and that can be accompanied by a warranty at least equal to that of a new product. The main steps of a remanufacturing process are summarised in Fig. 3.3. "*The production of remanufactured goods depends upon the supply of end-of-life finished goods for disassembly and remanufacture. The length and complexity of the remanufacturing process varies, with companies performing one or a number of different operations: storage and transport, disassembly, cleaning, inspection, trading of components, replacing worn parts, product restoration, testing and distribution*" [12].

Two main types of remanufacturing firms may be involved in the process described in Fig. 3.3:

- original equipment manufacturers (OEMs), that manufacture their own products and have the capability to collect them at end of life and the facility to remanufacture. In some cases they can propose contracts based on the formula "product-service". In any case they are interested in boosting forms of design for disassembly and of advanced logistics (traceability of products, programmes of periodical remanufacturing, modalities of deconstruction and collection, protocols of warranty, etc.);
- non-OEMs remanufacturers, namely remanufacturers that source products and core at the end of life either directly or in some cases through brokers. They can also be distributor-type organisations, selling OEM products and also providing repair, refurbishment and remanufacturing services.

The ERN project [12], sponsored under Horizon 2020, investigates all the main economic sectors and underlines that, despite of being a key strategy within the circular economy, remanufacturing is still "*an undervalued part of the industrial landscape and an under-recognised, sustainable industry*" and differently from Europe, "*USA and China, already have some common vision and strategy for remanufacturing while European remanufacturing could lose competitiveness against these more organised economies. As a result, there is a real need for a European-level solution to encourage remanufacturing throughout Europe*". The ERN report estimates the current level of remanufacturing activity within the EU and focuses on the most advanced sectors. As regards the construction sectors, only the sector of (office) furniture has been analyzed, even if it is smaller compared to the other ones that have

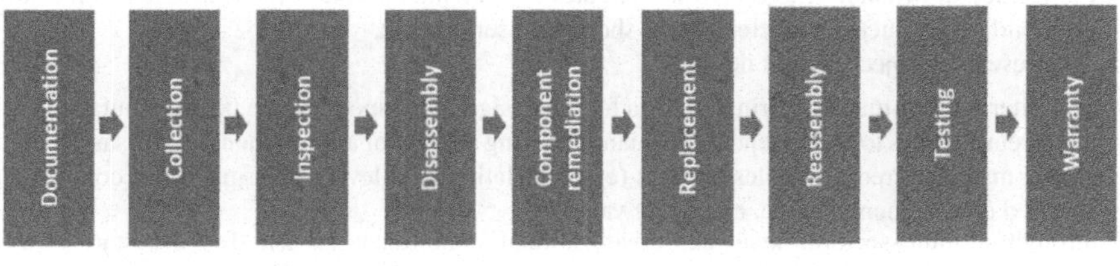

Fig. 3.3 Main steps of the process of remanufacturing. *Source* EC ERN (2015) [12]

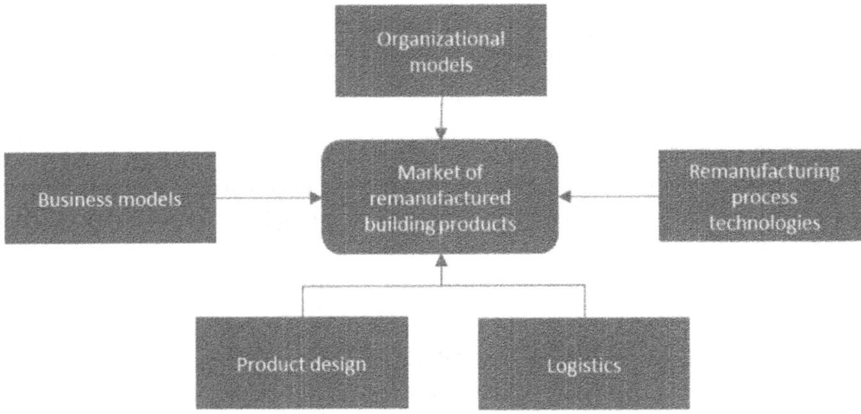

Fig. 3.4 Key areas for the development of a market of remanufactured building products

been investigated (aerospace, automotive, heavy duty and off-road (HDOR) equipment, electronic and electrical equipment (EEE), machinery and medical equipment). Surely the construction sector is lagging behind other sectors, but the lesson learned from other, more advanced, sectors might be very useful for the analysis of the key areas (Fig. 3.4) and conditions (drivers and barriers) for the development of a market of remanufactured building products.

"*Remanufacturing occurs across a wide range of industrial sectors, but is particularly attractive to industries that produce capital-intensive, durable products with relatively long product life cycles. However, these factors alone are not enough to enable remanufacturing to take place. The economic viability of remanufacture can depend on a number of factors including: how dispersed the product is; how easy it is to locate and retrieve; the ease with which it can be disassembled, diagnosed and remediated; the rate of technology and product performance change; knowledge and skills related to manufacturing; and legislative changes*" [12].

In the field of the construction sector, the attention of possible remanufactures[6] should be focused on short term components [20] with a service life lower than

[6]In this field, the research RENetTA, recently funded by Fondazione Cariplo, is being developed in Politecnico di Milano. The reserch investigates the potential second life or the extension of the life of building products, characterized by short-term service life.

The research project aims at defining:

- key criteria for re-manufacturing: criteria for the design of components "to be re-manufactured"; interpretative keys to understand the re-manufacturing attitude of a component to be disassembled and re-manufactured; obstacles/barriers (e.g. legislative) and levers (e.g. guarantee conditions, certified environmental value, economic value);
- through a multi-sectoral analysis: organizational conditions; criteria for the start of re-manufacturing processes; levers for the launch of successful re-manufacturing processes from an economic and environmental point of view; economic, environmental and social benefits to be used as levers for the start up of re-manufacturing processes;
- rules to support the re-manufacturing processes: relationship rules (organizational, procedural, etc.); assembly, disassembly and processing procedures in relation to various technical elements

15 years (interiors, services, equipment, furnishings and fittings which frequently became potential waste that can be converted into new resources), placed in tertiary buildings[7] (offices, temporary shops, retail, etc.), characterized by recurrent renovations due to recurrent activities of buying and selling and to the accelerated obsolescence. Accelerated obsolescence of internal components, usually dry-disassembled, and furnitures, is also typical in temporary exhibitions and fairs, that request recurrent interventions of disassembling/renewal/transformation.

3.3 Information Supporting Circularity

The construction works, considered as a source,[8] can provide a large amount of materials[9] to be recycled (Table 3.4) and of components to be reused/remanufactured both within the construction sector and within other manufacturing sectors. In Table 3.5 the virtuous example of the German building sector shows the self supply potential of building sector. But, at the same time the construction sector, as a potential receiver, can use recycled material and products deriving from the manufacturing sector in a cross sectoral market of secondary materials and reprocessed components. Between the drivers, above discussed, for this kind of market, surely information plays a fundamental role.s

Information is fundamental in the various stages of the processes of generation of secondary materials and components and its role may be interpreted according to various reading keys. A first reading key may be the one regarding information-based instruments and their role in disseminating knowledge about the measures for improving material resource efficiency and in boosting specific initiatives to close material loops in a circular economy. An overview of this perspective is provided by the EEA

(interior walls, finishes, floors, ceilings, external skin, windows frames); quality procedures, standards and methods for defining the characteristics of components easy to be re-manufactured; methods for exchanging materials and products; methods for sharing information and communication protocols; procedures of application of LC-based indicators aiming at identifying the environmentally and economically more effective strategies to be adopted in re-manufacturing processes (e.g. definition of the maximum advantageous physical distance between disassembly site and remanufacturer) [20].

[7]There is a direct relationship between the number of trades and the number of internal components/fittings renewals, caused by the change of the owners and, therefore, by a change of their needs and requests.

[8]See the project BAMB (Buildings As Material Banks) that deals with the development of a building pass. The aim of the project is the reduction or elimination of construction and demolition waste. In the process, buildings are regarded as material banks (http://www.bamb2020.eu).

[9]Commission Decision 2000/532/EC [21].

Table 3.4 Construction and demolition waste

ELoW code	ELoW label	EWC-Stat code
17	Construction and demolition wastes (including excavated soil from contaminated sites)	
17 01	Concrete, bricks, tiles and ceramics	
17 01 01	Concrete	W121
17 01 02	Bricks	W121
17 01 03	Tiles and ceramics	W121
17 01 06*	Mixtures of, or separate fractions of concrete, bricks, tiles and ceramics containing hazardous substances	W121
17 01 07	Mixtures of concrete, bricks, tiles and ceramics other than those mentioned in 17 01 06	W121
17 02	Wood, glass and plastic	
17 02 01	Wood	W075
17 02 02	Glass	W071
17 02 03	Plastic	W074
17 02 04*	Glass, plastic and wood containing or contaminated with hazardous substances	W121
17 03	Bituminous mixtures, coal tar and tarred products	
17 03 01*	Bituminous mixtures containing coal tar	W121
17 03 02	Bituminous mixtures other than those mentioned in 17 03 01	W121
17 03 03*	Coal tar and tarred products	W121
17 04	Metals (including their alloys)	
17 04 01	Copper, bronze, brass	W062
17 04 02	Aluminium	W062
17 04 03	Lead	W062
17 04 04	Zinc	W062
17 04 05	Iron and steel	W061
17 04 06	Tin	W062
17 04 07	Mixed metals	W063
17 04 09*	Metal waste contaminated with hazardous substances	W102
17 04 10*	Cables containing oil, coal tar and other hazardous substances	W102
17 04 11	Cables other than those mentioned in 17 04 10	W062
17 05	Soil (including excavated soil from contaminated sites), stones and dredging spoil	
17 05 03*	Soil and stones containing hazardous substances	W126
17 05 04	Soil and stones other than those mentioned in 17 05 03	W126
17 05 05*	Dredging spoil containing hazardous substances	W127
17 05 06	Dredging spoil other than those mentioned in 17 05 05	W127

(continued)

Table 3.4 (continued)

ELoW code	ELoW label	EWC-Stat code
17 05 07*	Track ballast containing hazardous substances	W121
17 05 08	Track ballast other than those mentioned in 17 05 07	W121
17 06	Insulation materials and asbestos-containing construction materials	
17 06 01*	Insulation materials containing asbestos	W12B
17 06 03*	Other insulation materials consisting of or containing hazardous substances	W121
17 06 04	Insulation materials other than those mentioned in 17 06 01 and 17 06 03	W121
17 06 05*	Construction materials containing asbestos	W12B
17 08	Gypsum-based construction material	
17 08 01*	Gypsum-based construction materials contaminated with hazardous substances	W121
17 08 02	Gypsum-based construction materials other than those mentioned in 17 08 01	W121
17 09	Other construction and demolition wastes	
17 09 01*	Construction and demolition wastes containing mercury	W121
17 09 02*	Construction and demolition wastes containing PCB (for example PCB- containing sealants, PCB-containing resin-based floorings, PCB-containing sealed glazing units, PCB-containing capacitors)	W077
17 09 03*	Other construction and demolition wastes (including mixed wastes) containing hazardous substances	W121

Sources Commission Decision 2000/532 [21]; ELoW codes and corresponding EWC-Stat Codes [22]
*No data are available

and ETC/WMGE[10] analytical report [3], that describes the findings from the analysis of information provided by countries, reviews national approaches to material resource efficiency and explores similarities and differences in policy responses. The study brings out the shared awareness for European countries towards information-based instruments. Responding to the question, put during the process of analysis,

[10]The analytical report was prepared by the EEA together with Eionet countries and the European Topic Centre on Waste and Materials in a Green Economy (ETC/WMGE). It presents an overview of findings from the analysis of information provided by European 32 countries, reviewing national approaches to material resource efficiency and exploring similarities and differences in policy responses. The analysis is based on the information provided by participating EEA member countries through the Eionet National Reference Centres on Resource Efficient Economy (NRCs) and the National Focal Points (NFPs). The main focus of the report is on policies and initiatives for material resources. The scope includes material flows entering or leaving an economy (biomass, non-metallic minerals, metal ores and fossil energy materials) as well as secondary (waste–derived) raw materials. Also within the scope are the transformations that materials undergo throughout their full life cycle, and initiatives to close material loops in the context of a circular economy.

Table 3.5 Self-supply potential with secondary construction materials within the German building sector for selected materials

		National usage		Building sector national usage		Potential secondary materials through construction waste		Self supply potential of building sector**
Material	Year	Mio. t/a	t/p/a	Mio. t/a	t/p/a	Mio. t/a	t/p/a	(%)
Mineral raw materials	2013	540	6,70	450	5,58	78,8	0,98	18%
Sand and gravel	2013	236	2,93	211	2,62	*		
Special sand and—sands	2013	9,7	0,12	3,5	0,04	*		
Natural stone	2013	207	2,57	196	2,43	*		
Limestone and dolomite	2013	23,7	0,29	7,1	0,09	*		
Gypsum nd anhydrite	2013	4,5	0,06	3,9	0,05	0,6	0,01	15%
Clay and kaolin	2013	13,3	0,16	3	0,04	*		
Brick clay	2013	11,3	0,14	11,3	0,14	5,1	0,06	45%
Recycling material	2013	66,2	0,82	66,2	0,82	*		
Mineral products								
Concrete	2014	110	1,36	110	1,36	21,9	0,27	20%
Cement	2014	27,3	0,34	27,3	0,34	0	0,0	0%
Glass	2015	2,9	0,04	1,9	0,02	0,27	0,003	14%
Metals								
Steel	2015	44,4	0,55	13,8	0,171	6,5	0,081	47%
Aluminium	2015	3,1	0,04	0,5	0,006	0,2	0,002	40%
Zinc	2015	0,64	0,01	0,5	0,006	0,025	0,0003	5%
Copper	2014	1,47	0,02	0,7	0,009	0,067	0,0008	10%
Lead	2014	0,36	0,00	0,036	0,0004	0,015	0,0002	42%

Source Heinrich [23]
*No data are available
**No system losses are considered

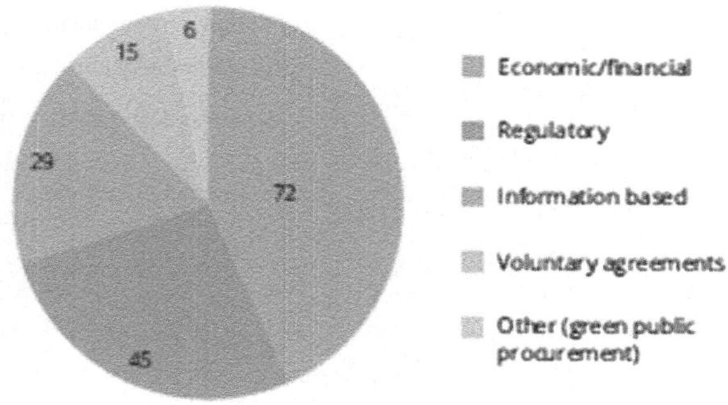

Fig. 3.5 The most important policy instruments for material resource efficiency, by number of mentions in country responses. *Source* EEA [3]

on which instruments are most important for material resource efficiency, countries reported 166 examples, among which: economic and financial instruments (72 reports) were mentioned most often, regulatory (45) and information-based instruments (29) followed in second and third place and voluntary agreements were the least mentioned group of instruments (15 reports) (Fig. 3.5).

Information-based instruments include communication and information campaigns, technical support schemes and eco-audits, training and education, or various eco-labels and awards to raise public awareness and increase the visibility of an issue. According to the EEA report [3], information-based instruments were reported by 18 countries. Information-based instruments for eco-labelling (5 reports) and for increasing awareness of the circular economy paradigm, especially for small and medium enterprises), were most commonly reported. For instance, in Flanders (Belgium), there are several information-based instruments that aim to support companies in becoming more aware of their use of materials, and to demonstrate to companies how they can become more sustainable and efficient. Estonia supports activities for raising awareness in companies, training resource specialists or auditors for resource-efficient solutions. Some countries target support for SMEs. Several countries (Belgium, Bulgaria, the Czech Republic, Estonia, France and Germany), boost specialized mechanisms such as industrial symbiosis or develop tools in order to help companies become more aware of their use of materials and identify options for becoming more resource efficient.

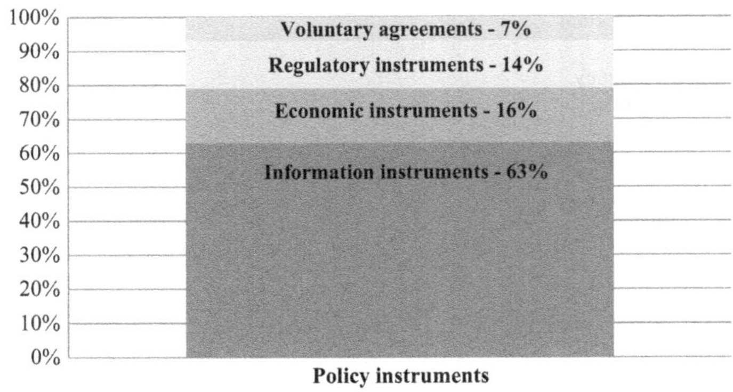

Fig. 3.6 Distribution of policy instruments in European waste prevention programmes. *Source* EEA [24]

This trend is confirmed also by another EEA study [24], dealing[11] with the review of the waste prevention programmes in Europe. It shows that the prevailing type of policy instruments are information instruments (Fig. 3.6) with 63% of all those listed across 27 waste prevention programmes, followed by economic and regulatory instruments accounting for 16 and 14%, respectively. Voluntary agreements are the least represented category, comprising about 7% of the total mix.

A second, no less important, reading key regarding information for circular economy is that related to building products across their whole life cycle. An example of this perspective can be found in the study, developed by the Rijkswaterstaat—Water, Verkeer en Leefomgeving National Institute for Public Health and the Environment [2], and regarding the kind of information that can describe the building products across their whole life cycle and allow stakeholders to communicate and interact and to share knowledge. The study explores the relevance of 'circular economy' for the Netherlands' construction sector and supports the view that the commissioning party—especially public authority—should have an important role in the development of good practices of information collection and knowledge sharing: "*Little information is shared between producing and processing companies. This is primarily a task for the companies that participate in the circular economy themselves. The*

[11] The EEA (2015) study "Waste prevention in Europe—the status in 2014" reviews by the waste prevention programmes in Europe. The review process covers programmes in the 28 European Union (EU) Member States and the three European Free Trade Association (EFTA) countries, namely Iceland, Liechtenstein and Norway.

The investigated waste prevention programmes contain a wide variety of policy instruments, but these can be grouped into four main types:

- information instruments, including labels, awareness-raising campaigns and pilot projects;
- regulatory instruments, setting binding standards and norms;
- economic instruments, including tax incentives, green public procurement and direct subventions;
- voluntary agreements, with clear and measurable targets, mainly initiated by business associations.

government, as commissioning party, is however a link in the construction chain (or cycle) and quite often long-time owner of the materials used in construction. As the commissioning party, the government is rarely aware of the materials that are used and of the stocks of material under its control (...) Mainly, the government—as a user—is the link between the suppliers and producers on the one hand and demolition and recycling companies on the other. Because information about materials must be kept as long as the materials are in use, this is a clear task for the government as the commissioning party in construction" [2]. The subject of information, constantly linked to the building products over time, represents a big challenge for recycling and reuse of CDW. At this regard the materials passport [2] aims to allow information to travel with the product itself through time. The development of the materials passport is still in its infancy. BIM (Building Information Model) [2] can be a possible design tool, useful for specifying and sharing, between different operators, the circular requirements for secondary materials, and the related information necessary for describing them. The BIM 3D design tool already contains an extensive database of materials. The addition of specific circular information, that is relevant for the following link in the chain is already being investigated. The greatest challenge is possibly how to store and keep such information accessible so that it can be usefully employed during the demolition—50 to 100 years later. At this aim Akinade et al. [25] propose a functionality framework for BIM-based Design for Deconstruction tools, Akanbi et al. [26] develop a BIM-based whole-life Performance Estimator (BWPE) for appraising the salvage performance of structural components of buildings right from the design stage, Bilal et al. [27] present a proposal for a BIM-enabled building waste analysis (BWA).

A complete set of data, able to describe characteristics of materials and performances of building products along their lifespan may allow to: assess the feasibility of recycling and reuse, facilitate the pre-demolition audits, reduce of uncertainty about quantity and quality of materials and components and the perception of the potential health risk for workers, create the conditions for a marketplace.

Considering existing buildings, "*evidence from fieldwork observations, waste documentation, and interviews indicates that the generators of unwanted components effectively decide their fate, and a failure to identify components in advance, uncertainty over usefulness, the perception of cost and programme risk in reclamation, and the preferential order of the waste hi*erarchy mean that the decision to discard to waste management goes unchallenged" [4]. For this reason it is important "*to capture timely information about existing building components to be discarded, make this information visible to a wide community, and determine usefulness by focusing creativity already present in the industry on an exhaustive examination of component reusability and upcyclability. The existing literature thus agrees on the need for better information about existing building components and marketplaces for their exchange. The assumption is that this will allow contractors and demolition contractors to understand where there is demand for items they would otherwise discard, and initiate reclamation and resale*" [4].

The lack of information about building components to be deconstructed lead to a lack of confidence in the quality of CD recycled materials as well as uncertainty about the potential health risk for workers using recycled CD materials. This lack of confidence reduces and restricts the demand for recycled and reused materials and components, which in turn inhibits the development of CDW market.

The EU Construction & Demolition Waste Management Protocol [28] represents a proposal, articulated in many objectives and actions (Fig. 3.7), aiming to overcome many barriers, including those connected with low levels of knowledge.

The Protocol fits within the Construction 2020 strategy, as well as the Communication on Resource Efficiency Opportunities in the Building Sector, and is also part of the Circular Economy Package that the European Commission has presented. The Protocol is addressed to various stakeholders: industry practitioners; construction

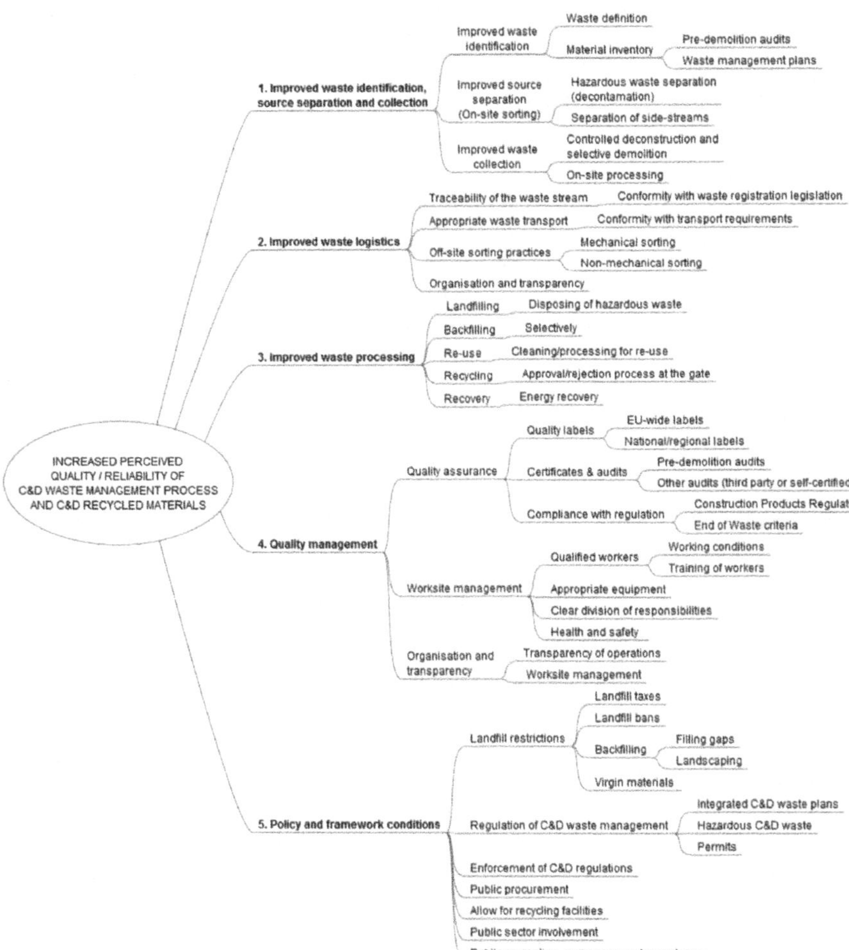

Fig. 3.7 The objectives and actions tree of the EU CD Waste Management Protocol. *Source* EU Construction & Demolition Waste Management Protocol [28]

sector (including renovation companies and demolition contractors), construction product manufacturers, waste treatment, transport and logistics as well as recycling companies; Public authorities at local, regional, national and EU levels; Quality certification bodies for buildings and infrastructure; clients of CD recycled materials.

The Protocol deals with the generation and collection of unambiguous information in various actions of its breakdown structure (Fig. 3.7), especially in the aspects related to waste identification. In particular the Protocol states that:

- "*A pre-demolition audit (or waste management audit) is to be carried out before any renovation or demolition project and for any materials to be re-used or recycled, as well as for hazardous waste. It helps to identify the C&D waste generated, implement proper deconstruction, and to specify dismantling and demolition practices. The actions based on the audit will ensure the safety of workers and lead to an increase in the quality and quantity of recycled products. It will also help to increase the amount of materials to be re-used close to or at the construction site. Additionally, the implementation of these audits can help clients with setting performance levels for demolition contractors, support a site-specific waste management plan, demonstrate environmental credentials, increase material and labour efficiency, reduce waste and maximise profit*";
- *public authorities should decide upon the threshold for pre-demolition audits*[12];
- for collecting information it is necessary to identify: all waste materials that will be generated during the demolition[13] with specification of the quantity, the quality and location in the building or civil infrastructure;

[12]As an example, in Austria two limits for pre-demolition audits exist: 100 tons and 3.500 m^3 of estimated CD waste produced.

[13]To give two examples:

- the French regulation for construction and building projects specifies how to identify waste from demolition and refurbishment of buildings. The buildings concerned are those, industrial or commercial, with a surface area of more than 1000 m^2 for each floor or farm, that have been exposed to hazardous substances. The works concern the reconstruction and/or demolition of a major part of the structure of the building. The entity in charge must carry out the identification, before applying for the demolition permit or before accepting estimates for contracting, by specifying: the characteristics, the quantity and the location of material and waste, the modalities of management, which materials are re-used on site, recovered or eliminated. At the end of the works, the contracting authority writes an assessment of works indicating the nature and the amount of material actually re-used on site and that of waste that is recovered or eliminated. The contracting entity sends the form to the French Environment and Energy Management Agency which presents a yearly report to the Ministry in charge of construction. See: www.legifrance.gouv.fr/affichTexte.do?cidTexte=JORFTEXT000025145228;
- the Dutch certification scheme for demolition processes (BRL SVMS-007) is a voluntary tool to improve the quality of the demolition process. Customers who prescribe to this certification scheme of procurement and tendering are assured of environmentally and safe demolition on site. The scheme is controlled by third parties and the Council of Accreditation. The certified demolition process follows four steps: pre-demolition audit; waste management plan; execution; final report. See: www.veiligslopen.nl/en/home.

- which materials must (mandatory) be separated at source (such as hazardous waste), which materials can/cannot be re-used or recycled;
- how the waste (non-hazardous and hazardous) will be managed and the recycling possibilities;
- a pre-demolition audit should consider the local markets for CD waste and reused and recycled materials, including the available capacity of recycling installations;
- the necessary presence of qualified experts for the pre-demolition audits, with appropriate skills about building materials, building techniques, building history, demolition techniques, possible markets waste treatment and processing.

Finally, a further reading key of information for boosting the market of secondary products can be defined by considering the point of view of the platforms for matching demand and offer of materials and products upstream and downstream of the recycling/reuse processes. The platforms may have different configurations and aims, like the ones described in Chap. 5, but all they have in common the objective of connecting various stakeholders, sharing information and creating trades. In Chap. 5 the subject of the waste platform is discussed in depth and a framework of platform is proposed.

References

1. EEA (2018) Report No 4/2018, Waste prevention in Europe—policies, status and trends in reuse in 2017
2. RWS (2015) Circular economy in the Dutch construction sector. A perspective for the market and government, final report
3. EEA (2016) More from less—material resource efficiency in Europe 2015 overview of policies, instruments and targets in 32 countries, EEA Report No 10/2016
4. Rose C, Stegemann J (2018) From waste management to component management in the construction industry. Sustainability 10(1):229
5. Hobbs G, Adams K (2017) Reuse of building products and materials–barriers and opportunities. In: International HISER conference on advances in recycling and management of construction and demolition waste 21–23 June 2017, Delft University of Technology, Delft, The Netherlands
6. Bougrain F, Laurenceau S (2017) Reuse of building components: an economic analysis. In: International HISER conference on advances in recycling and management of construction and demolition waste 21–23 June 2017, Delft University of Technology, Delft, The Netherlands
7. COM (2015) 614, Communication from the commission to the European parliament, the council, the European economic and social committee and the committee of the regions. Closing the loop—an EU action plan for the Circular Economy
8. EEA (2016) Circular economy in Europe—developing the knowledge base, EEA Report No 2/2016, European environment agency. EEA, 2017, Circular by design—Products in the circular economy, EEA Report No 6/2017, European Environment Agency
9. Directives (EU) 2018/851 of the European parliament and of the council of 30 May 2018 amending Directive 2008/98/EC on waste
10. Directives 2008/98/EC on waste, 94/62/EC on packaging and packaging waste, 1999/31/EC on the landfill of waste, 2000/53/EC on end-of-life vehicles, 2006/66/EC on batteries and accumulators and waste batteries and accumulators, and 2012/19/EU on waste electrical and electronic equipment

11. EC (2017) Resource efficient use of mixed wastes. Improving management of construction and demolition waste. Final report
12. EC ERN (2015) Remanufacturing market study, report
13. EN 13306: 2010, Maintenance—maintenance terminology
14. APRA Europe (2012) Remanufacturing terminology: remanufacturing term guideline
15. BS 8887: 2009, Design for manufacture, assembly, disassembly and end-of-life processing (MADE) Part 2: Terms and definitions
16. EMF (2013) Towards the circular economy, Ellen MacArthur Foundation, Cowes, United Kingdom
17. Triple Win (2014) The economic, social and environmental case for remanufacturing, report
18. CLEPA Reman Definition & Process (2011)
19. ISO 9000:2005, Quality management systems—fundamentals and vocabulary
20. Talamo C, Lavagna M, Monticelli C, Atta N, Giorgi S, Viscuso S Re-NetTA. Re-manufacturing networks for tertiary architectures. In: Della Torre S, Cattaneo S, Lenzi C, Zanelli A (eds) The regeneration of the built environment from a circular economy perspective, Springer, In press
21. Commission Decision 2000/532/EC of 3 May 2000 replacing Decision 94/3/EC establishing a list of wastes pursuant to Article 1(a) of Council Directive 75/442/EEC on waste and Council Decision 94/904/EC establishing a list of hazardous waste pursuant to Article 1(4) of Council Directive 91/689/EEC on hazardous waste (notified under document number C(2000) 1147) (Text with EEA relevance)
22. ELoW codes and corresponding EWC-Stat Codes (2017), Resource Efficient Use of Mixed Wastes – Improving management of construction and demolition waste—Final report 2017
23. Heinrich MA (2017) Material flows of the German building sector. In: International HISER conference on advances in recycling and management of construction and demolition waste 21–23 June 2017, Delft University of Technology, Delft, The Netherlands, pp. 302–305
24. EEA (2015) Waste prevention in Europe—the status in 2014, Report No 6/2015
25. Akinade OO, Oyedele LO, Omoteso K, Ajayi SO, Bilal M, Owolabi HA, Looney JH (2017) BIM-based deconstruction tool: towards essential functionalities. Int J Sustain Built Environ 6(1):260–271
26. Akanbi LA, Oyedele LO, Akinade OO, Ajayi AO, Delgado MD, Bilal M, Bello SA (2018) Salvaging building materials in a circular economy: a BIM-based whole-life performance estimator. Resour Conserv Recycl 129:175–186
27. Bilal M, Oyedele LO, Qadir J, Munir K, Akinade OO, Ajayi SO, Owolabi HA (2015) Analysis of critical features and evaluation of BIM software: towards a plug-in for construction waste minimization using big data. Int J Sustain Build Technol Urban Dev 6(4):211–228
28. EU Construction & Demolition Waste Management Protocol (2016)

Chapter 4
Waste Up-Cycling in EU Co-funded Projects

Abstract The European Union in order to promote and stimulate forms of innovation aimed at the preservation of the environment and to practices of circular economy, have been significantly contributing for years to finance projects aimed at both product innovation and process innovation. The most famous systems of financing are the Life programs, the Horizon 2020 and the CIP; among these, focusing the attention on the building sector, several successful projects are oriented to enhance and reconsider scraps and waste as a secondary raw material. The chapter examines some significant projects and highlights the main trends through some interpretation keys (type of innovation implemented, main production sectors involved, types of waste most used, etc.). The knowledge deriving from this study may help to understand the new perspectives in the recycling field and what are the main innovative strategies pursued by the participating companies. The analysis carried out provides both a method of analysis and a tool for the definition of strategies for the replication and improvement of new industrial initiatives.

Keywords Circular economy · Up-cycling · Waste · Valorization · Life programme · CIP · Horizon 2020

4.1 The EU Environmental Funding Systems: Life, CIP, COSME, Horizon 2020

4.1.1 Life Programme

The European Union promotes and finances projects and initiatives aimed at environmental improvement in support of the need to promote adaptation and productive systems to the new demands of environmental sustainability, but also in response to the attention given by national governments, companies and citizens to the "green" products market. The first funding plan was launched in 1992, the financial

This chapter is authored by Marco Migliore, Politecnico di Milano, Department of Architecture, Built Environment and Construction Engineering, Milan, Italy. e-mail: marco.migliore@polimi.it.

M. Migliore et al., *Strategies for Circular Economy and Cross-sectoral Exchanges for Sustainable Building Products*, Springer Tracts in Civil Engineering,
https://doi.org/10.1007/978-3-030-30318-1_4

Table 4.1 Scheduling of life projects funded from 1992 to 2017, with indications of the budget made available by EU (data processed by www.ec.europa.eu)

Life I_1992–1995								
1992	**1993**	**1994**	**1995**	**–**	**–**	**–**	**Projects** [n]	**Budget** [bil€]
105	144	245	237	–	–	–	731	400
Life II_1996–1999								
1996	**1997**	**1998**	**1999**	**–**	**–**	**–**	**Projects** [n]	**Budget** [bil€]
181	192	221	281	–	–	–	875	450
Life III_2000–2004 e Life III_2005–2006								
2000	**2001**	**2002**	**2003**	**2004**	**2005**	**2006**	**Projects** [n]	**Budget** [bil€]
239	0	216	208	204	–	–	867	640
–	–	–	–	–	160	130	290	317,2
Life +_2007–2013								
2007	**2008**	**2009**	**2010**	**2011**	**2012**	**2013**	**Projects** [n]	**Budget** [bil€]
141	194	211	182	202	249	228	1.408	2.143
Life_2014–2020								
2014	**2015**	**2016**	**2017**	**2018**	**2019**	**2020**	**Projects** [n]	**Budget** [bil€]
144	148	139	142	–	–	–	573	1796
–	–	–	–	n.a.	–	–	–	1658
Total							**4.744**	**7.973,2**
Life_2021–2027								
2021	**2022**	**2023**	**2024**	**2025**	**2026**	**2027**	**Projects** [n]	**Budget** [bil€]
–	–	–	–	–	–	–	–	5.450

instrument promoted was the LIFE[1] program. The main purpose of the LIFE Program is to contribute to the implementation, updating and development of EU environmental policy and legislation by co-financing pilot projects or trials with added value at European level. The Life program was started in 1992 [1] and since then five re-financings have been started, of which four have been completed and one is still in progress. Through the LIFE program more than 4700 projects have been financed, with an investment in the protection and safeguarding of the environment and climate of almost 8 billion euros (Table 4.1).

The results achieved by the financed projects have encouraged over the years the refinancing of the program and the constant increase[2] in the budget [1–6] made

[1]Life is a Community financial instrument established in 1992, following the issue of two important Community Directives, 79/409/EEC ("Birds" Directive) and 92/43/EEC ("Habitat" Directive), with the Regulation EEC n. 1973 to support projects in the EU and in some candidate countries and neighboring.

[2]Over the years the budget made available for the Life program has grown significantly. In the first funding (Life I 1992–1995) the budget made available for the four-year period was around 400 million euros, a condition that was kept constant until the third loan (Life III 2000–2004 and 2005–2006) which had received more funds. Starting from the fourth refinancing (Life+2007–2013)

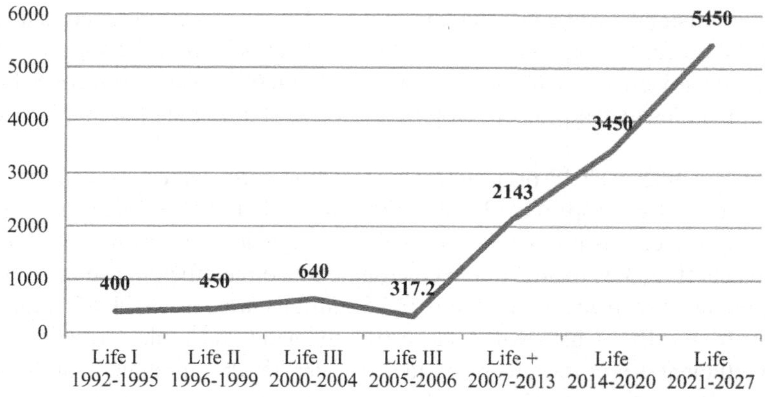

Fig. 4.1 Evaluation of the increase in funds made available for the Life program, from the beginning until the next refinancing (data processed by www.ec.europa.eu)

available (Fig. 4.1). For the 2014–2020 period, financial availability was significantly more consistent than in other years. Janez Potočnik, European Commissioner for Environment, in launching the call for proposals for 2014–2020, said that through these contributions and projects we will contribute to "*achieve sustainable growth through investments in an economy that is efficient in the use of resources and will allow Member States and local authorities to implement plans and strategies in key areas such as nature, waste, air and water*" [7].

For the sixth funding (2021–2027) the European Commission [8] declared the proposal to increase the investment to 5.43 billion euros, with an increase of 60% compared to the financing in progress (increase of about 2 billion EUR). Miguel Arias Cañete, Climate Action and Energy Commissioner, added: "*A stronger LIFE programme will play an important role in expanding investments in climate action and clean energy across Europe. By continuing to support climate change mitigation and adaptation, LIFE will also continue to help the EU deliver on its climate goals and commitments under the Paris Agreement and the United Nations Sustainable Development Goals*" [8].

The long-term success of the Life program undoubtedly derives from its theme addressed exclusively to the environment, but above all to the role played to guarantee greater solidarity and a better sharing of responsibilities for the preservation of the common environmental and climate good in the Union. It is currently divided into thematic areas, which respond to the main needs in environmental terms, these areas are supported differently with each refinancing, taking into account the environmental situation. In the current financing plan, two sub-programs are distinguished: Environment and Climate Action. The sub-program for the Environment includes three priority areas: environment and resource efficiency, nature and biodiversity

we chose to invest much more, in fact the budget increase has been doubled. In the fifth refinancing (Life 2014–2020) the budget was further increased by 60% and for the next refinancing (Life 2021–2037) a further 60% increase was announced.

and environmental governance and information. The sub-program for climate action includes three areas of action: climate change mitigation, climate change adaptation and climate governance and information. Each of these areas has objectives [9] that become features of the projects proposed to the Commission, which allocates the funds.

In the funding of 2017 projects for 430 million euros were financed, the funds made available by Europe were 242.8 million euros, about 56% of the total (Fig. 4.2).

Among these, it is possible to observe a considerable distribution of funds for the projects (Fig. 4.3) related to the environment and the efficient use of resources, which obtained 55 loans for a total of 163.5 million euros, of which 82, 4 allocated directly by the EU, which will finance five thematic areas: air, environment and health, efficient use of resources, waste and water. Furthermore, it should be noted that to facilitate the transition towards a more circular economy in Europe, the 20 projects on the efficient use of resources alone will mobilize 43.8 million euros, with an increase of 15% compared to last year. Approximately € 14.9 million will be used to improve air quality.

The second priority area in terms of project and fund allocation is the one related to Nature and Biodiversity, which received 40 projects and 153 million euros to which the EU will contribute 97.5 million. The third priority area in terms of project assignment is the one related to climate change adaptation, which has been awarded 17 projects and 44.2 million euros to which the EU will contribute with 22.9 million euros. This funding will cover projects related to six different themes: ecosystem-based adaptation, health and well-being, adaptation of mountainous/island areas focused on the agricultural sector, urban adaptation/planning, vulnerability assessment/adaptation strategies, and water (including flood management, coastal areas and desertification). The fourth priority area in terms of project assignment is the one related to the governance and information component that has been awarded 15

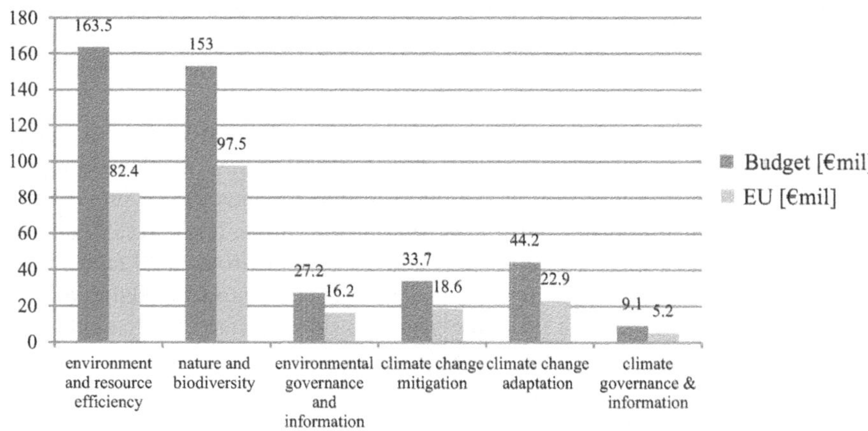

Fig. 4.2 Distribution of funding for the year 2017 with respect to the priority areas of the two sub-programs (data processed by www.ec.europa.eu)

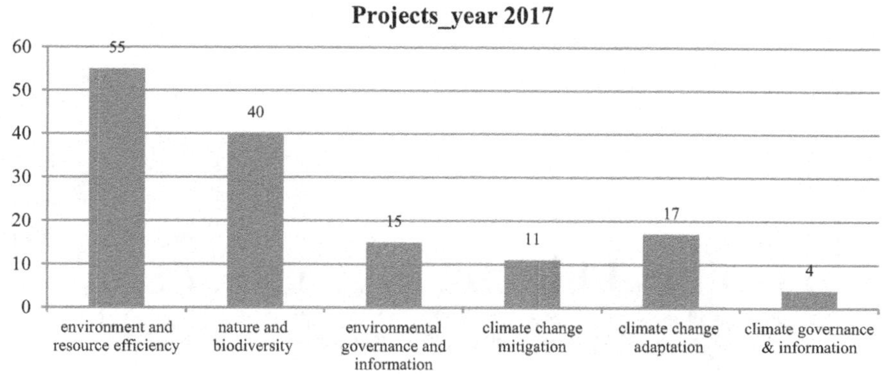

Fig. 4.3 Distribution of projects for the year 2017 with respect to the priority areas of the two sub-programs (data processed by www.ec.europa.eu)

projects and 27.2 million euros to which the EU will contribute 16.2 million euros, the projects in this area will be aimed at raising awareness of environmental issues. The fifth priority area in terms of project assignment is the one related to the mitigation of climate change, which received the assignment of 11 projects and 33.72 million euros to which the EU will contribute 18.6 million euros. This funding covers best practice projects, pilot projects and experimentations in three thematic areas: industry, accounting/reporting of greenhouse gas emissions, and land use/forestry/agriculture. The area that has received the least projects is the one related to the governance and information component on the climate, which has obtained 4 projects and 9.1 million euro to which the EU will contribute with 5.2 million.

An interesting information about Life projects is the one concerning the distribution of funds with respect to the participating nations. The constant presence of some nations emerges (Fig. 4.4) in all editions of the program, with a significant number of assigned projects. In particular it is possible to observe how Spain and Italy have received the major number of assigned projects.

4.1.2 CIP

The "Entrepreneurship and Innovation Program" EIP [10], set up to support and sustain innovation between Small and Medium-sized Enterprises (SMEs) and to promote the improvement of their competitiveness, is part of the broader "Competitive and Innovation Framework Program "CIP, aimed at fostering a growing competitiveness of companies operating in the European Community. The first CIP program ended in 2013 and it had a six-year application period, from 2008 to 2013. Table 4.2 summarizes a comparison between the 2008, 2009, 2010, 2011, 2012 and 2013 calls regarding: the growing request for shareholdings, the amounts of funds and the percentage of small and medium-sized enterprises [11].

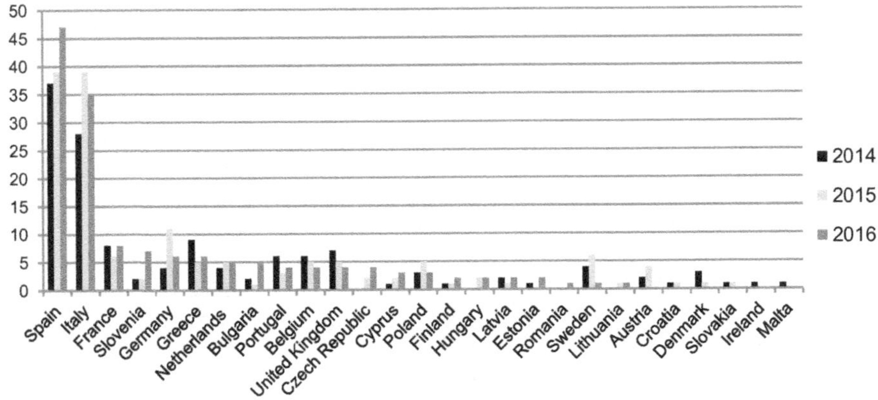

Fig. 4.4 Distribuzione dei progetti per stati nel periodo compreso tra il 2014 e il 2016 (dati elaborati da www.ec.europa.eu)

Table 4.2 Summary table of calls from 2008 to 2013 indicating the projects financed, the number of participants and the funds allocated (data from EACI [11])

	Call 2008	Call 2009	Call 2010	Call 2011	Call 2012	Call 2013
Number of proposal	134	202	287	279	284	461
Number of participants	444	614	895	860	916	1518
Requested founding[a] (M€)	110	150	264	199	196	309
Average request founding (k€)	830	770	921	712	690	695
SME (%)	74	70	66	67	67	64

[a] *To avoid an excessive distortion of the budgetary estimations, the following proposals were not accounted: sixteen proposals with no EU funding requested, one proposal that requested more than 50% for the project and one proposal with an unreasonably high EU funding request* [11]

During this period € 3621 million have been invested in projects involving the following three operational programs:

- EIP
  Entrepreneurship and Innovation Programme.
  The Entrepreneurship and Innovation Program has been proposed to support innovation in SMEs, focusing on some particular aspects: facilitating the access to available "financial instruments"; favoring the creation of organized European networks of companies; supporting innovation policies and encouraging "eco-innovative" initiatives that can contribute to the reduction of environmental impacts, to the reduction of pollution and to a better control of the use of natural resources. For this programme, a budget of 2,17 billion euro has been reserved, including 430

million to promote eco-innovation, aiming at facilitating SMEs access to finance, at better integrating the existing networks of business support services (EuroInfoCentres and Innovation Relay Centres) and at supporting innovation activities (INNOVA, Pro-Inno, etc). More than 1 billion euro will be devoted to boost the highly successful financial instruments managed by the European Investment Fund (EIF), which co-invests in venture capital funds (covering early stage and expansion stage), and provides co-guarantees on loans [12].

- ICT-PSP
  Information Communication Technologies Policy Support Programme.
  The Information and Communication Technologies Policy Support Program aims to stimulate the wide dissemination of innovative services based on ICT and digital exploitation in Europe by citizens, governments and businesses (particular attention is reserved to SMEs). The main objective of the program is to guide the dissemination of ICT in areas of public interest, addressing EU changes towards a low carbon economy, for this programme, a budget of 730 million euro has been reserved [12].
- IEE
  Intelligent Energy Europe Programme.
  The Intelligent Energy Program, with approximately 730 million euro, has contributed to the achievement of ambitious goals for the management of climate change and energy use. The program supported projects and best practices such as: favoring the design and construction of buildings that can lead to a 50% saving in energy consumption compared to traditional buildings, supporting the production of electricity from renewable sources throughout Europe and supporting European cities in converting their public mobility into cleaner and more efficient systems [12].

Each of these programs has specific objectives, all designed to contribute to the competitiveness of businesses and to support their innovative capacity in the sectors in which they operate, in ITCs and in renewable energy. Within the Entrepreneurship and Innovation Program, several projects were financed (Fig. 4.5) related to some specific sub-categories, in particular: materials recycling and recycling processes, sustainable building products, food and drink sector, water efficiency, treatment and distribution and greening business.

There is considerable interest in the recycling and waste valorization sector, which can be seen across projects involving the construction sector. Furthermore, from the general evaluation of the program [13], some significant data emerge:

- estimated global benefits of 1.6 billion euro, generated by the application and systemization of the selected projects;
- the widespread and growing interest in experimental approaches to recycling and "green" innovation;
- wide participation in small and medium-sized projects, the percentage is significantly higher than in large companies;

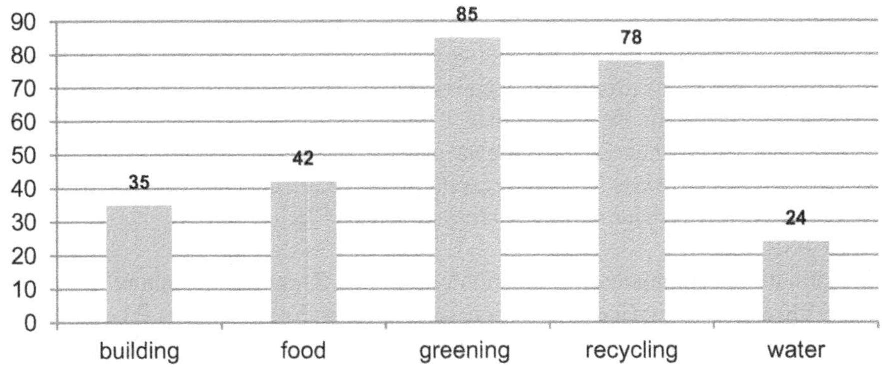

Fig. 4.5 Distribution of projects funded by the EIP fund on the five thematic areas: materials recycling and recycling processes, sustainable building products, food and drink sector, water efficiency, treatment and distribution and greening business (data from www.ec.europa.eu)

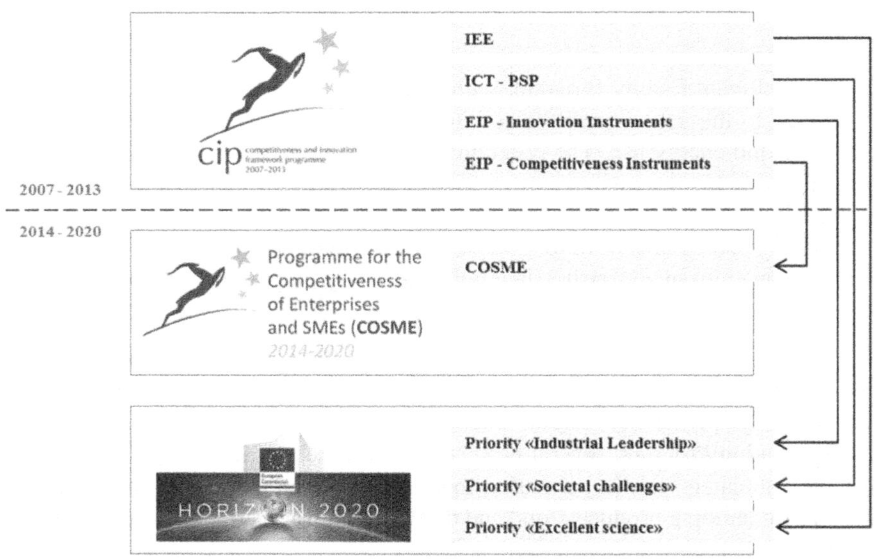

Fig. 4.6 Temporal succession of financing initiatives

- large participation by nations that have suffered most from the economic crisis (Italy and Spain are the main beneficiaries with € 4.9 billion a year the first and € 5.1 billion the second).

The CIP program alone has managed to finance about 250,000 small and medium-sized enterprises (which would hardly have had credit), creating 220,000 jobs [14]. On the basis of this data, at the end of the CIP program (Fig. 4.6), the COSME Program (Competitiveness of Enterprises and Small and Medium-Sized Enterprises) [15] was introduced. This is a program (2014–2020) focused on the competitiveness of companies with particular attention to SMEs. The COSME program has a budget of 2.5 billion euros [16] which will be useful to guarantee loans and venture capital actions, sufficient to mobilize estimated capital of 25 billion euros. Two financial instruments have also been established: the Loan Guarantee Facility[3] (LGF) e la Equity Facility for Growth[4] (EFG), which will aim to protect and safeguard investors and investments, as well as to facilitate financial transactions.

4.1.3 Horizon 2020

Horizon 2020 [16] is the Framework Program of the European Union (EU) for research and innovation covering the period from 2014 to 2020. The Framework Programs have a seven-year duration and they represent the main instrument with which the Union Europe (EU) funds research in Europe.

Horizon 2020, differs from other programs (Life and CIP) because it unifies three previous programs (2007–2013) in a single financial instrument aimed at supporting research, innovation and technological development: the Seventh Framework Program[5] (7FP), the Framework Program for Competitiveness and Innovation (CIP) and the European Institute of Innovation and Technology[6] (EIT). Its elaboration was started in 2011, when the EU Heads of State and Government invited the European Commission, for the 2014–2020 period, to integrate into a common strategic framework the various tools dedicated to support research and innovation. The Commission has thus initiated a broad consultation that has involved all the main players in the field of research and has led to the establishment of the Program.

---

[3]The *Loan Guarantee Facility* (LGF) is a tool created ad hoc by the EU for Small and Medium Enterprises, which in many cases, without the appropriate guarantees, could not aspire to obtain a loan. In these terms, this tool intends to remedy these shortcomings by making available: guarantees for debt financing (up to 50% of the loan) and securitization of loan portfolios granted to SMEs, for amounts up to 150,000 euros.

[4]The *Equity Facility for Growth* (EFG) will instead be the instrument for making investments in risk capital funds, which will make investments in SMEs in a state of expansion and/or growth.

[5]The Seventh Framework Program for Research and Technological Development (FP7) is the main tool through which the European Union funds research in Europe. The Seventh Framework Program, which was in force from 2007 to 2013, is the legitimate successor of the Sixth Framework Program and it is the result of years of consultations with scientific community, research institutes, decision-making bodies and other interested parties. Since their launch in 1984, the framework programs have played a leading role in multidisciplinary research and cooperation activities, in Europe and beyond. FP7 has fulfilled this function, but it has been more extensive and more comprehensive than in previous framework programs. The program was in force from 2007 to 2013 and it had a budget of 53.2 billion euros over seven years, the largest so far provided for these programs.

[6]The European Institute for Innovation and Technology (better known through the acronym EIT) is an agency of the European Union created in 2008 with the aim of identifying, co-financing and coordinating the activity of specific "communities of knowledge and innovation".

The purpose of Horizon 2020 is to support research and innovation: one of the five main objectives[7] to which Europe 2020 aims, the European Union strategy for the decade 2010–2020. In line with this strategy, H2020 aims to contribute to the creation of a society based on knowledge and innovation, oriented towards the great priorities defined by the European Agenda for 2020: smart, sustainable and inclusive growth (Fig. 4.7).

Conceived as a tool capable of supporting and guiding economic development and above all creating new jobs, Horizon 2020 represents the main financial tool aimed at strengthening the European Research Area, that is, the creation of a common area where researchers, scientific and technological knowledge can circulate freely and implement the Innovation Union (the Europe 2020 flagship initiative aimed at promoting European global competitiveness, favoring the creation of partnerships for innovation, strengthening research initiatives and simplification of the administrative procedure for access to funds). Horizon 2020 has a budget of almost 80 billion euros over seven years (Fig. 4.8). The total amount is over 30%, compared to the previous program.

Horizon 2020 is built around three priorities, each including specific objectives: Excellent Science, Industrial Leadership and Societal Challenges.[8]

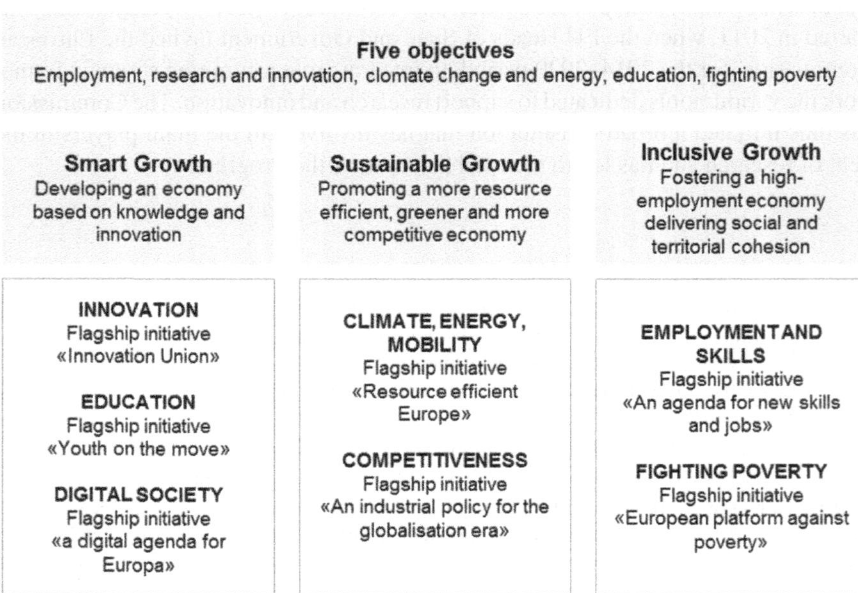

Fig. 4.7 Schematization of the five objectives to which Horizon 2020 aims and of the three major priorities of growth

[7]Employment, research and innovation, climate change and energy, education and the fight against poverty, are the five main objectives that Horizon 2020 aims to achieve.

[8]https://ec.europa.eu/programmes/horizon2020/en/h2020-sections.

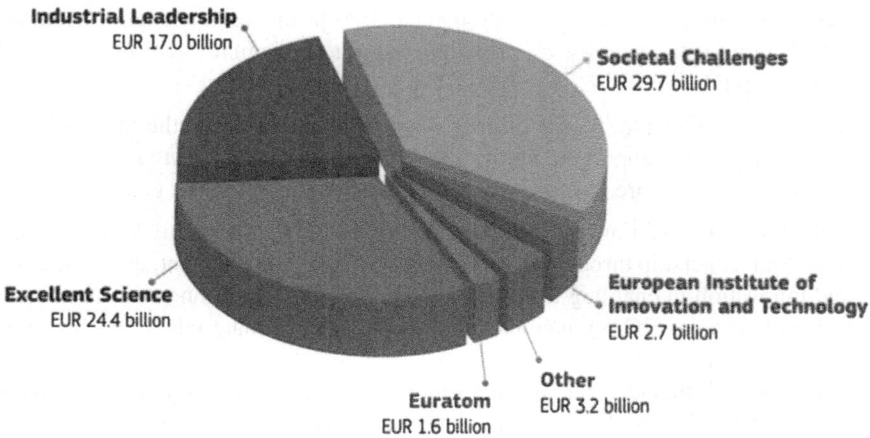

Fig. 4.8 Distribution of resources by priority areas (image taken from https://www.researchitaly.it/en/horizon-2020/)

Excellent Science.[9] This priority is oriented towards the reinforcement and strengthening of Europe in the basic science. For this reason it is built around four specific goals:

- European Research Council (ERC),[10] to encourage frontier research in Europe by supporting individual talented researchers and their teams;
- Future and Emerging Technologies (FETs),[11] in order to support research on radically new advanced technologies with a high-risk pioneering ideas that can innovate the actual productive system;
- Marie Skłodowska-Curie[12] actions, to strengthen skills, training and career development of researchers by encouraging cross-border and cross-sector mobility;

[9]https://ec.europa.eu/programmes/horizon2020/en/h2020-sections.

[10]The mission of the European Research Council is to encourage the highest quality research in Europe through competitive funding and to support research across all fields, on the basis of scientific excellence. The ERC complements other funding activities in Europe such as those of the national research funding agencies, and it represents a flagship component of Horizon 2020, the European Union's Research Framework Programme for 2014–2020 (from www.erc.europa.eu).

[11]FET actions are expected to initiate radically new lines of technology through unexplored collaborations between advanced multidisciplinary science and cutting-edge engineering. It will help Europe to quickly seize its leadership in those promising technological areas able to renew the basis for future European competitiveness and growth, and that can make a difference for society in the decades to come. Under Horizon 2020, FET actions had a provisional budget of 2696 million euro (from www.ec.europa.eu).

[12]The Marie Skłodowska-Curie actions (MSCA) provide grants for all the stages of researchers' careers—from doctoral candidates to highly experienced researchers—and they encourage transnational, inter-sectoral and interdisciplinary mobility. The MSCA enable research-focused organizations (e.g. universities, research centers and companies) to host talented foreign researchers and to create strategic partnerships with leading institutions worldwide. The MSCA aim to support researchers in gaining the necessary skills and the international experience for a successful career, either in the public or the private sector. The programme responds to the main challenges faced

- Research infrastructures,[13] to strengthen European research infrastructures, including e-infrastructures, by providing value to their innovative potential and human capital.

Industrial Leadership.[14] This priority is oriented to accelerate the technological development and to support the innovation of SMEs and their growth in the international context. Therefore, this priority is built around three specific goals:

- Leadership in enabling industrial technologies. The goal is to boost Europe industrial leadership through research, technological development, demonstration and innovation in enabling technologies, including information and communication technologies (ICTs), nanotechnologies, advanced materials, biotechnology, advanced manufacturing.
- Access to risk finance. The aim is to support enterprises in increasing—through specific financing tools, including loans and guarantees—their access to risk finance to invest in research and innovation.
- Innovation in SMEs. The goal is to encourage different types of innovation in small- and medium-sized enterprises, in particular those with high growth and internationalization potential, by encouraging the creation of a supportive ecosystem.

Societal challenge.[15] This priority addresses future social issues. It is built around seven challenges:

- Health, demographic change and wellbeing—to improve the lifelong health and well-being of all;
- Food security, sustainable agriculture, forestry, marine, maritime and inland water and bio-economy—to secure sufficient supplies of safe, healthy and high quality food by developing sustainable and efficient production systems;
- Secure, clean and efficient energy—to make the transition to a reliable, affordable, publicly accepted, sustainable and competitive energy system;
- Smart, green and integrated transport—to achieve a European transport system that is resource-efficient, climate- and environmentally-friendly and safe;
- Climate action, environment, resource efficiency and raw materials—to achieve a resource- and water-efficient and climate change resilient economy and society;
- Europe in a changing world—inclusive, innovative and reflective societies—to foster a greater understanding of ongoing changes and to provide solutions for a sustainable growth at the social and economic levels;

---

by researchers, offering them attractive working conditions and the opportunity to move between academic institutions and similar (from www.ec.europa.eu).

[13]The European approach to research infrastructures has made a remarkable progress in recent years with the implementation of the European Strategy Forum on Research Infrastructures (ESFRI) roadmap, integrating and opening national research facilities and developing e-infrastructures underpinning a digital European Research Area. The networks of research infrastructures across Europe strengthen its human capital base by providing world-class training for a new generation of researchers and engineers and promoting interdisciplinary collaboration (from www.ec.europa.eu).

[14]https://ec.europa.eu/programmes/horizon2020/en/h2020-sections.

[15]https://ec.europa.eu/programmes/horizon2020/en/h2020-sections.

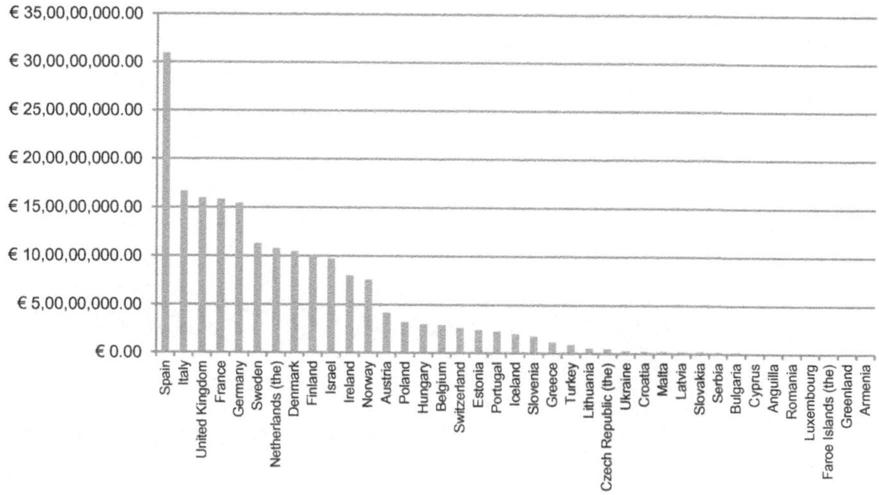

Fig. 4.9 Comparison of the financial attributions made to the various EU countries through the Horizon 2020 programme (data processed from http://sme.easme-web.eu)

- Secure societies—protecting freedom and security of Europe and its citizens—to foster secure European societies and to address global threats, while strengthening the European culture of freedom and justice.

Additional action.[16] Moreover, H2020 provides funds for the following actions:

- Spreading excellence and widening participation in order to ensure that the benefits of an innovation-led economy are maximized and widely distributed across the European Union;
- Science with and for the society, with the objective of building an effective cooperation between science and society, recruiting new talent for science and pairing scientific excellence with social awareness and responsibility;
- Cross activities, concerning industry for 2020 in circular economy, Internet of Things, smart and sustainable cities;
- Fast-track for innovation, a pilot initiative focused on the promotion of market-related innovation activities;
- European Institute for innovation and technology, to support the EU organization to promote Europe's competitiveness;
- Euratom, an international organization established under the complementary programme for research and education in the field of nuclear energy;
- Cyber-physical systems, with the purpose of enhancing information technology systems that interact with the physical context in which they operate.

Figure 4.9 shows the distribution of funds provided to date by the Horizon 2020 programme, for EU member states. This distribution shows that Spain is one of the

[16] https://ec.europa.eu/programmes/horizon2020/en/h2020-sections.

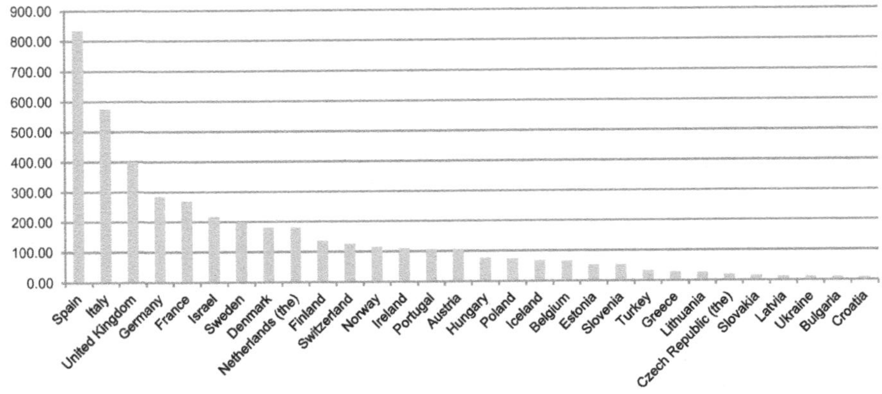

Fig. 4.10 Distribution by state of projects funded through the Horizon 2020 programme (data processed from http://sme.easme-web.eu)

main financed countries, with a budget exceeding € 300 million, Italy is in second place with € 160 million.

With respect to the number of projects funded, Fig. 4.10 shows the distribution by state of the projects assigned and started in the EU member states. Even in this case Spain is one of the main financed countries, with 834 projects assigned, Italy is in second place with 575 projects.

4.2 Trends of the Financed Projects

As part of a research funded by the "Fondazione Fratelli Confalonieri di Milano" entitled "Ri-Scarto", studies on funded research projects were conducted, paying particular attention to those concerning the up-cycling and valorization of waste. The aim of the research was to highlight good practices that led to environmental improvements in the supply chain, product or management procedures, so that they can be replicated in other production contexts. The study highlighted the type of actors who applied for European funding, the type of innovation implemented and the type of waste treated.

4.2.1 Life Projects on Waste Up-Cycling

The Life program is articulated into sub-programs, which finance projects related to the specific areas identified. In particular, the sub-program areas are:

- Nature and biodiversity
  Nature conservation projects (biodiversity, habitats and species) are a priority of the Life programme. For this reason, projects related to best practice, pilot and demonstration projects that contribute to the implementation of the EU's directives on birds and habitats—the EU's biodiversity strategy to 2020—and to the development, implementation and management of the Natura 2000 network, can receive an high percentage of co-founding. Typically, the selected projects receive a co-founding till 60%, but in some case the co-financing rate can be incremented to 75% (only if the half of the total costs of the estimated projects are used for actions to improve the conservation status of priority habitats or species listed in the EU's birds and habitats directives).
- Environment and resource efficiency
  Life programme co-finances projects related to the environmental sector. In particular, the priority areas are: air, chemicals, green and circular economy, industrial accidents, marine and coastal management, noise, soil, waste, water, and the urban environment. The Life programme supports initiatives for pilot and demonstration projects in order to test, develop and demonstrate management and industrial policy. In order to identify best practices and solutions, the Life programme provides support to the development and improvement of innovative technologies and to the evaluation and monitoring of EU's environmental policy. Technologies and solutions that are ready to be implemented in close-to-market conditions, on a commercial scale or on an industrial scale, are considered more valid than others and they are more easily approved. The selected projects receive a co-founding till 55%.
- Environmental governance and information
  The area related to governance and information, supports projects concerning topics as environmental training and capacity building, legislative compliance and enforcement, knowledge development and public and stakeholder participation. In order to increase awareness on environmental issues, the Life programme provides support tool to sharing information and disseminate projects, including cooperation platforms and platform for sharing knowledge on sustainable environmental solutions and practice. The selected projects receive a co-founding till 55%.

Within these areas are included projects that have been examined to identify specificities and trends related to the best management of natural resources and to the enhancement of waste materials (by-products and waste). The principle according to which they were selected concerns the presence within projects of strategies for the innovation in process and production management, and in some cases, innovation in information management practices. The selection of projects was made through the projects database[17] of Life programme by inserting as a reference period the period between 1992 and 2019. The first action carried out in the study of the projects

[17]The projects database of Life programme is available at the following link http://ec.europa.eu/environment/life/project/Projects/index.cfm and it groups together all Life projects funded from 1992 to today. Through the database it is possible to select projects using different reading keys (year, countries, topics, beneficiary, keywords, etc.).

connected with the topics of "waste" (Fig. 4.11) and "resources" (Fig. 4.12) was to quantify and categorize them by main areas of action. Compared to the "environment" program, in the case of waste it emerges that around 582 projects concern this topic and they are distributed heterogeneously in different categories (agricultural waste, bio-waste, C&D waste, ELV's and tyres, End-of-pipe land-filling, hazardous waste, industrial waste, medical waste, municipal waste, packaging and plastic waste, WEEE, waste recycling, waste reduction and waste use), which identify the areas of greatest interest. Among these areas, there are numerous projects concerning waste recycling, reduction and use—which represent around 45% of the total of the selected

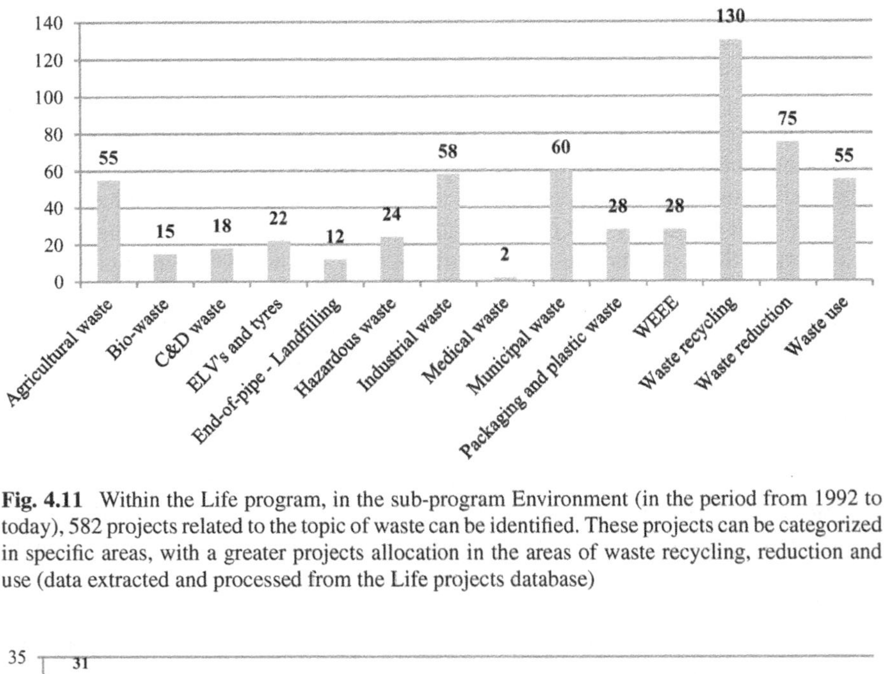

Fig. 4.11 Within the Life program, in the sub-program Environment (in the period from 1992 to today), 582 projects related to the topic of waste can be identified. These projects can be categorized in specific areas, with a greater projects allocation in the areas of waste recycling, reduction and use (data extracted and processed from the Life projects database)

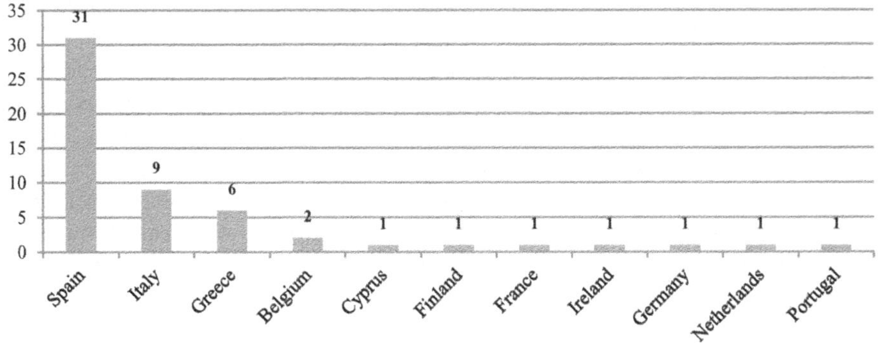

Fig. 4.12 Distribution by state of projects related to the "waste" issue in the agricultural sector (data extracted and processed from the Life projects database)

projects—and many projects concerning the agricultural sector, the industrial sector and the urban (city) sector.

For the agricultural sector, which counts over 9% of the total number of funded projects, it is possible to highlight some particular aspects:

- 31 projects (Fig. 4.12) have been assigned to Spain, about 56% of the total projects funded in this area, achieving a record among the European states that deal with this topic. This is a very interesting fact because, on one hand, it shows a parallelism with the high agricultural production of Spain and the quantity of soil used for this purpose and, on the other hand, it highlights a national interest in the transaction of the agricultural sector towards a more sustainable reality.
- A second very important fact on the subject is that Spain, unlike other states (France, Germany and the United Kingdom), has a very high number of companies linked to the agricultural sector. This represents a condition that describes a scenario of small productive realities—and not of large companies as in other states—which nevertheless expresses much more interest in innovating then other European contexts.
- With respect to the type of projects, there is a high interest in the production of bio-fuels, in the up-cycling of waste as fertilizers and in the up-cycling of waste for the production of green packaging.

For the industrial waste sector, which accounts almost 10% of the total number of funded projects, it is possible to highlight some particular aspects:

- 23 projects were assigned to Spain (Fig. 4.13), about 53% of the total of projects funded in this area. This is a record among European states that are interested in the topic.
- The product sectors involved are very heterogeneous, the common line is that all the projects concern the up-cycling of scraps/waste with intrinsic value, with different purposes: production of new materials, integration of second raw material in the supply chains and cross-sectoral transfer of scraps. In addition, there are also some

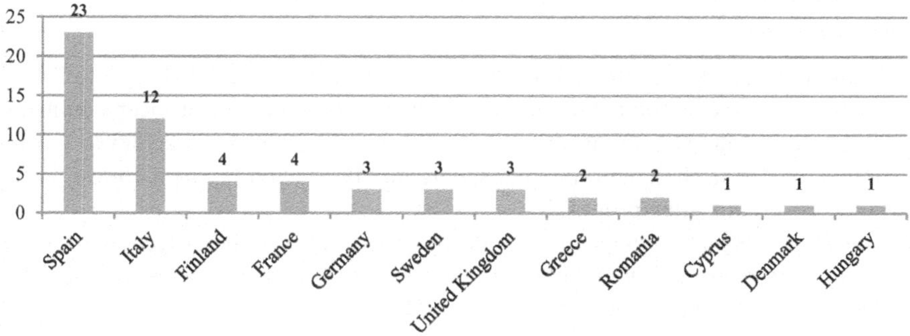

Fig. 4.13 Distribution by state of projects related to the topic of industrial waste (data extracted and processed from the Life projects database)

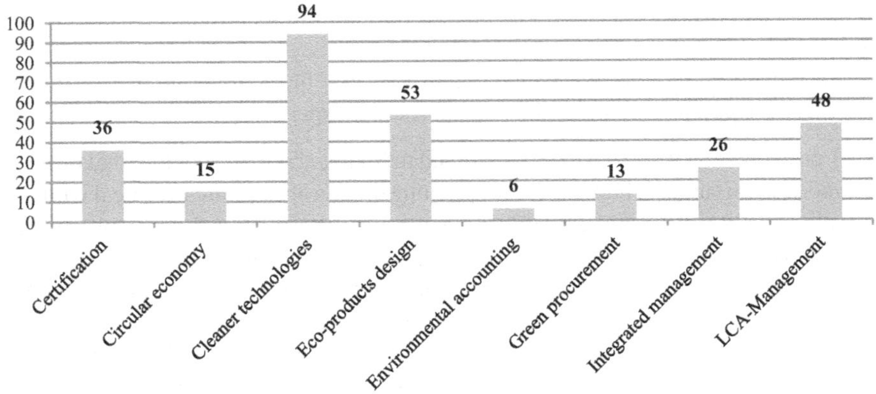

Fig. 4.14 Within the Life program, in the sub-program Environment, (in the period from 1992 to today) 291 projects related to the theme of environmental management can be identified. These projects can be categorized in specific areas, with a greater allocation of projects in the areas of cleaner technologies, eco-products design and LCA-management (data extracted and processed from the Life projects database)

projects that aim to transform production or management systems, regardless of scraps/waste treatment.

In the case of "environmental managements" it emerges that, with respect to the "environment" program, around 291 projects (Fig. 4.14) concern this issue and they are distributed heterogeneously in different categories (certification, circular economy and value chains, cleaner technologies, eco-products design, environmental accounting, green procurement, integrated management, Life Cycle Assessment-Management), which identify the areas of greatest interest. These include numerous projects that concern the development of cleaner technologies, the implementation of an eco-products design and of a Life Cycle Assessment Management.

Projects involving the development of cleaner technologies represent more than 32% of the total number of funded projects and it is possible to highlight some particular aspects:

- Spain and Italy have obtained funding for 59 projects (Fig. 4.15), about 62% of the total of projects funded in this area, with a record among European countries. This trend emerges from all the samples, these two states have achieved a truly unique record in the environmental sphere and they stand out as the most active in terms of innovative proposals.
- The proposals implemented through the projects cover very heterogeneous sectors (construction, agriculture, chemistry, etc.) and they provide interesting ideas for possible replication of the initiatives carried out.

A common aspect of all the initiatives is that of favoring projects whose aim is a widespread improvement of the environmental condition, both within the production chains and in the surrounding territories. In fact, it emerges that all the initiatives

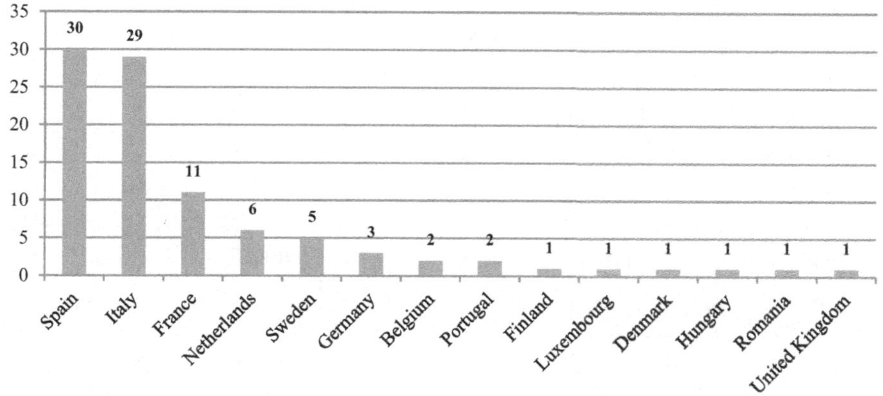

Fig. 4.15 Distribution by state of projects related to the topic of cleaner technologies (data extracted and processed from the Life projects database)

face problems with a spirit of innovation that have positive repercussions both on the production chains (waste production, waste disposal, efficiency of the supply chain, innovation of the production process, etc.) and on the territory (management environmental emergency deriving from industrial or intensive agricultural activity, management of urban settlements, etc.).

A further aspect that has been evaluated regarding the assignment of Life projects is the type of beneficiary, useful information to outline the profile of the actors who are most interested in the issues addressed by the Life programme. The sample used for the evaluation of main actors is that of the entire 1992–2019 Life program (4194 projects) and the identified typologies of actors are: development agencies, intergovernmental bodies, international enterprises, large enterprises, local authorities, mix enterprises, NGO foundations, national authorities, park-reserve authorities, professional organizations, public enterprises, regional authorities, research institutions, SMEs, training centers and universities. From the evaluation carried out (Fig. 4.16), it emerges that the sample of beneficiary is very heterogeneous, however most of the projects are attributed to applicants who are not directly connected with the market, but mostly to beneficiaries who have a close connection with the territory (regional authority, local authority, research institution, etc.). The only exception, which also represents the record in assignments, is that of NGO foundations, which have obtained funding for 767 projects, 18% of the internal number of projects funded. The motivation lies in the fact that most of these projects concern the Natura sub-program, moreover the reference NGOs represent extended authorities across the different EU states, with deep roots in the territories.

Lastly, with respect to what has already been reported, a specific focus was conducted on the building supply chain, highlighting projects that directly concern the construction sector (production of building materials, construction phase, building

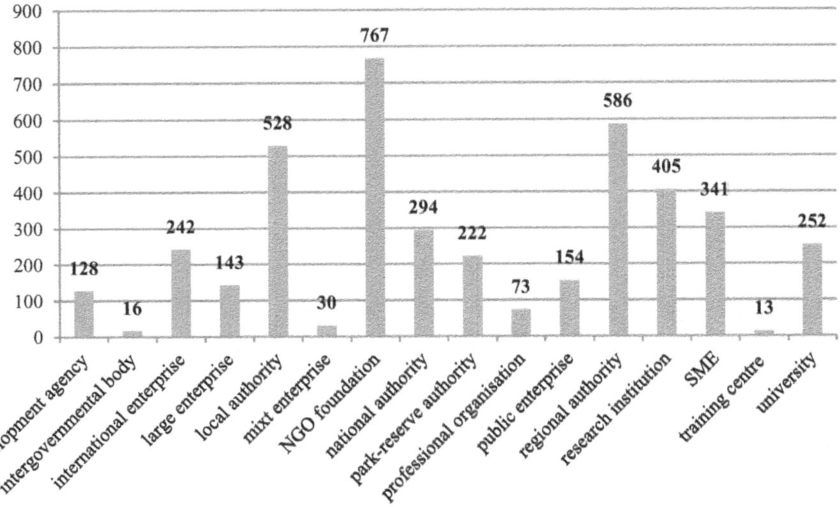

Fig. 4.16 Distribution by type of beneficiary of the projects (data extracted and processed from the Life projects database)

disposal). The aim is to highlight the increasing efforts to make this chain environmentally efficient and to underline how the initiatives cover almost all the production sectors connected to the construction industry. Approximately 100 projects, with close connection to the construction sector, were selected from the database. These projects have been analyzed to identify which are the main operating trends and which are the main addressed emergencies. Specifically, five main categories of strategies based on the prevention or reuse/recycling of building waste have been identified:

a. Reduction of non-exploitable waste.

In this context, numerous projects can be mentioned, since they collect one of the main objectives of many companies and/or trade associations. Waste reduction is one of the first prerogatives to favor a growing sustainability. By way of example, two projects are here introduced: LIFE08 ENV/E/000126[18] and LIFE+12 ENV/IT/000436.[19]

The first project "Eco-stone. Sustainable system implementation for natural stone production and use",[20] promoted by an association of professionals, has developed a procedure for the best management of the natural stone production chain with the aims of: reducing waste production (through process optimization); reducing energy consumption (through substantial modifications of processing systems); make

[18]LIFE08/ENV/E/126 [17], Eco-stone. Sustainable system implementation for natural stone production and use, Final Report, 2013 (available at https://ec.europa.eu/easme/en/life).

[19]LIFE+12/ENV/IT/000436 [18], LIFE Sustainable Mission—Test 1.0 of chemical industry for global sustainable organization as industrial total symbiosis and low energy and water, Project Technical Report, 2015 (available at https://ec.europa.eu/easme/en/life).

[20]cfr 19.

a sustainable product. The premise to this project was the awareness that: to produce about 200 m^2 of stone slabs for floors it will be necessary to extract at least 100 tons of natural stone from the quarry, with an overall process efficiency of 7%.

The second project "Sustainable Mission—Test 1.0 of chemical industry for global sustainable organization as industrial total symbiosis and low energy and water",[21] promoted by a large enterprise, has worked on the development of an innovative method able to eliminate the production of solid and liquid waste of the production process of ceramic tiles, using recycled raw materials (up to 70% of recycled waste with a high percentage of glass elements) to create tiles suitable for different fields of use. In this way, not only have the production waste been reduced, but also the ceramic glass waste from other production contexts have also been enhanced, being up-cycled in very high percentages. Another strong point of this project regards the improvement of the production process to reduce energy consumption and consequent emissions.

b. Development of innovative products with high eco-sustainability value that contain recycled waste.

By way of example, two projects are here presented: LIFE+12 ENV/IT/000374[22] and LIFE05 ENV/DK/000158.[23]

The first project "Plastic Killer—Innovative plastic pollutants removal for efficient recycled wood panels production",[24] promoted by a SME, aims to meet the growing demand for recycled wood for the production of MDF (Medium-density fibreboard) panels. Currently the MDF panels are obtained mainly from virgin wood and only a small number of producers use post-consumer recycled wood (up to 10%, based on internal studies by Imal Pal Group). This is due to the fact that this process requires many cleaning steps, which are not always sufficient to completely remove the impurities also due to the current cleaning systems. Often the MDF panels obtained are of low quality and they do not comply with the EN 622-5 standard and with the EPF standards for what regards the conditions of supply of recycled wood. The project aims at the birth of a new generation of more sustainable MDF panels (up to 60% recycled content) compliant with both the EN 622-5 standard and the EPF standards.

The second project "Recycling/symbiosis—waste and recycling and symbiosis in Stone wool production",[25] similarly to the other projects in this category, aims to test the possible reuse of processing waste as a component for the realization of mineral wool for insulation, through the reuse of aluminum processing waste, carbogrit and sawdust from natural stone. Typically the scraps considered in this

---

[21] cfr 20.

[22] LIFE+12/ENV/IT/000374 [19], Life Plastic Killer—Innovative plastic pollutants removal for efficient recycled wood panels production, Layman Report, 2017 (available at https://ec.europa.eu/easme/en/life).

[23] LIFE05/ENV/DK/000158 [20], RECYCLING/SYMBIOSIS—Waste and Sewage Recycling and Symbiosis in Stone Wool Production, Layman Report, 2008 (available at https://ec.europa.eu/easme/en/life).

[24] Cfr. 23.

[25] Cfr. 24.

project were destined for landfill, instead it was demonstrated—through the analyzes carried out—that their combination can lead to the creation of insulating material with excellent performance.

c. Optimization of production processes in order to facilitate recycling activities.

By way of example, two projects are here presented: LIFE08 ENV/E/000167[26] and LIFE+12 ENV/IT/000904.[27]

The first project "ICEJET—Ice Jet Environmental Technology Pilot Plant for Drastically Reducing Waste Produced by Abrasive Water Jet Cutting Techniques"[28] is based on the assumption that processing often contaminates waste to the point in which it inevitably turn into scraps for disposal in landfills, because it is not reusable. The project was born with the aim of designing a new production system able to characterize the waste, so that it can always be up-cycled.

The second project "FRELP. Full Recovery End Life Photovoltaic"[29] focuses on abandoned photovoltaic panels, in order to develop innovative technologies for the recovery of 100% of photovoltaic panels at the end of their life in an economically sustainable way. In particular, the project focused on the procedures and techniques to be adopted in order to ensure a total disassembly of the constituting materials of a panel.

d. Innovation in organizational models.

By way of example, two projects are here presented: LIFE+09 ENV/IT/000188[30] and LIFE+10 ENV/IT/000356.[31]

The first project "ECO-cluster—Environmental COoperation model for Cluster"[32] aims to develop a new industrial and territorial management system called ECO-cluster. The model developed aims to encourage a specific territory to take full control over the general environmental performance, through the shared effort of the (public or private) organizations that become part of the cluster. A territory involved in the implementation of the ECO-cluster model is facilitated in the creation of synergies between actors of the same cluster thanks to the opening towards a series of opportunities, including: increased competitiveness, diffusion and facilitation of innovation, optimized management of local issues.

---

[26] LIFE08/ENV/E/000167 [21], ICEJET—Ice jet environmental technology pilot plant for drastically reducing waste produced by abrasive water jet cutting techniques, Final Report: technical version, 2013 (available at https://ec.europa.eu/easme/en/life).

[27] LIFE+12/ENV/IT/000904 [22], LIFE FRELP—Full Recovery End-of-Life Photovoltaic Lyman Report, 2016 (available at https://ec.europa.eu/easme/en/life).

[28] Cfr. 27.

[29] Cfr. 28.

[30] LIFE+09/ENV/IT/000188 [23], ECO-CLUSTER—Environmental COoperation model for Cluster, Layman's Report, 2015 (available at https://ec.europa.eu/easme/en/life).

[31] LIFE+10/ENV/IT/000356 [24], Gy.Eco—Gyproc Eco-friendly, Report, 2016 (available at https://ec.europa.eu/easme/en/life).

[32] Cfr. 31.

The second project "Gy.Eco—Gyproc Eco-friendly"[33] was developed with the aim of creating a management and recovery system for waste coming from installation and after-sales activities of retailers and applicators operating in the field of dry systems. This project allows, on one hand, to contain the use of virgin raw material and, on the other hand, to achieve the total up-cycling of processing scraps. In addition, the scrap upturn service is provided directly by the dry-system manufacturer who, in this way, derives economic benefits.

e. Consumer awareness through knowledge dissemination.

The project LIFE+08 INF/IT/000312 "PROMISE—PROduct Main Impacts Sustainability through Eco-communication"[34] involves the implementation of a communication plan to raise awareness about environmental impacts of products as well as to inform and spread knowledge about sustainability tools. In particular, in this case, the strategy aims to contribute to environmental sustainability starting not from the product but from the environmental awareness of consumers.

4.2.2 *CIP Projects on Waste Upcycling*

Compared to the Life programme, projects funded with the CIP programme are numerically less but still interesting. Also in this case, as in the Life project financing programme, there is a wide spread of funding among nations, but there is a high concentration of projects in Italy, Spain, France and Germany. With reference to the thematic areas of the construction and recycling sector (Fig. 4.17), the trend remains and there is a greater interest in the recycling sector, which in some cases includes initiatives referred to the construction sector.

A specificity of the competitiveness and Innovation Framework Programme (CIP) is that it promotes projects that concern the first application or the replication on the market of eco-innovative techniques, products or processes, which contribute to reducing the environmental impact and to optimize the use of resources. Therefore, these are projects which—at the end of their course of study/experimentation—led to a systematic transformation of supply chain and production.

Both in the "buildings and construction" area and in the "recycling" area it is possible to trace experiences of considerable interest concerning the waste up-cycling.

[33]Cfr. 32.

[34]LIFE+08/INF/IT/000312 [25], PROMISE—Product Main Impacts Sustainability through Eco-communication, Technical Report, 2013 (available at https://ec.europa.eu/easme/en/life).

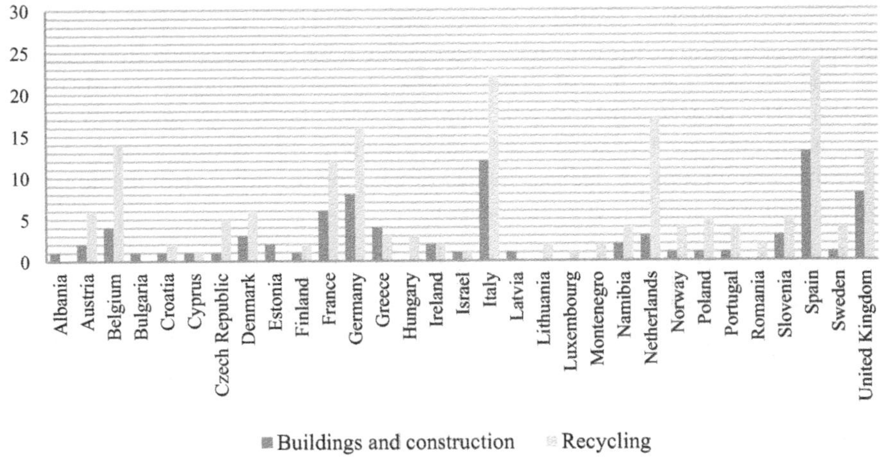

Fig. 4.17 Allocation of CIP funds by state to projects related to the construction sector and the recycling sector

In the first area the following projects are considerably relevant: "ECO/13/630249" [26],[35] "ECO/13/630185" [27][36] and "ECO/10/277233" [28].[37]

The first project "WOOL4BUILD"[38] was developed with the aim of improving the insulating performance of buildings through the use of building products deriving from the wool processing waste. The starting point of the project lies on the awareness that in this way it is possible to enhance up to 50.000 kg/year of wool by-products. It is not uncommon to find initiatives that involve the exploitation of waste and by-products as a second raw material for the production of insulating material. This procedure is made possible because almost all the insulating materials require a treatment of the input materials. In fact, the second project "Eco-innovative insulating thermal and acoustic panels made with recycled textile fibers (INSUL-ECO)"[39] also involves the production of insulating material starting from textile waste deriving from the treatment of out-of-service tires. The project involves Italy, France and Spain and it plans to manage the 120,000/150,000 tons of textile waste from tire recovery that are produced annually in Europe. The third project "Innovative recycled plastic based panels for building field. ECOPLASBRICK"[40] it starts from the awareness

[35]ECO/13/630249, Wool4build. Improved isolation material for eco-building based on natural wool (available at http://wool4build.com).

[36]ECO/13/630185, Insul-eco. Eco-innovative insulating thermal and acoustic panels made with recycled textile fibres (available at https://ec.europa.eu/environment/eco-innovation/projects/en/projects/insul-eco).

[37]ECO/10/277233, (2014), ECOPLASBRICK. Innovative recycled plastic based panels for building field, Layman Report, 2014 (available at http://ec.europa.eu/environment/eco-innovation/).

[38]Cfr. 36.

[39]Cfr. 37.

[40]Cfr. 38.

that the different polymers present in the mix of plastic-containing waste have a poor compatibility between them, which also implies a reduced mechanical resistance. This leads to an inevitable landfill disposal of this mix. Starting from this premise, the project proposes the use of mixed plastic as the core of a sandwich panel with a gres face-sheet (skins) on both sides.

In the second area, the projects "ECO/08/239071" [29][41] and "ECO/13/630426/WINCER" [30][42] should be mentioned. They involve the recycling of waste and by-products in sectors related to the construction supply chain.

The first project "Sludge free-process for the production of innovative natural stone-like obtained by micro-structuring of sintered tiles (NATSTOCER)"[43]—proposed by a large enterprise related to the ceramic sector—aims to industrialize a new production process able to reduce the use of high quantity of water and abrasive with silicon carbide as happen in the traditional production process. This new process involves important environmental improvements concerning the reduction of: use of virgin raw materials, water consumption and waste production. In particular, the process—with the use of air and raw material dust in the surface finishing processes of some materials—allows: the recycling of abrasive materials as a raw material (not contaminated materials) and the elimination of the high quantity of special waste namely the sludge deriving from the landfill processing with silicon carbide wheels. The second project "Waste synergy in the production of INnovative CERamic tiles (WINCER)"[44] aims to develop innovative ceramic tiles containing more than 70% of recycled materials from urban and industrial wastes. The specific objectives of the project are mainly related to: the recovery of the amount of soda lime glass cullet[45] waste that today is not re-introduced in glassware (about 30% of the total glass waste) and the improvement of environmental performance of the ceramic tiles sector by reducing CO_2 emissions, energy consumption and methane use.

[41]ECO/08/239071, NATSTOCER. Sludge free-process for the production of innovative natural stone-like obtained by micro-structuring of sintered tiles (available at https://ec.europa.eu/environment/eco-innovation/projects/en/projects/natstocer).

[42]ECO/13/630426, WINCER. Waste synergy in the production of INnovative CERamic tiles (available at https://ec.europa.eu/environment/eco-innovation/projects/en/projects/wincer).

[43]Cfr. 42.

[44]Cfr. 43.

[45]Packaging glass (soda lime based) comes from urban collection, it is separated, washed, purified and milled to be suitable for its reuse in the glass industry (glass with particle size 0.1–0.8 mm). The fraction that contains impurities and the finer fraction (below 100 μm) are not suitable for reuse use in the glass industry (about 30% of the total urban collection). This finer fraction is employed in WINCER tile formulation, as a substitution of feldspar flux.

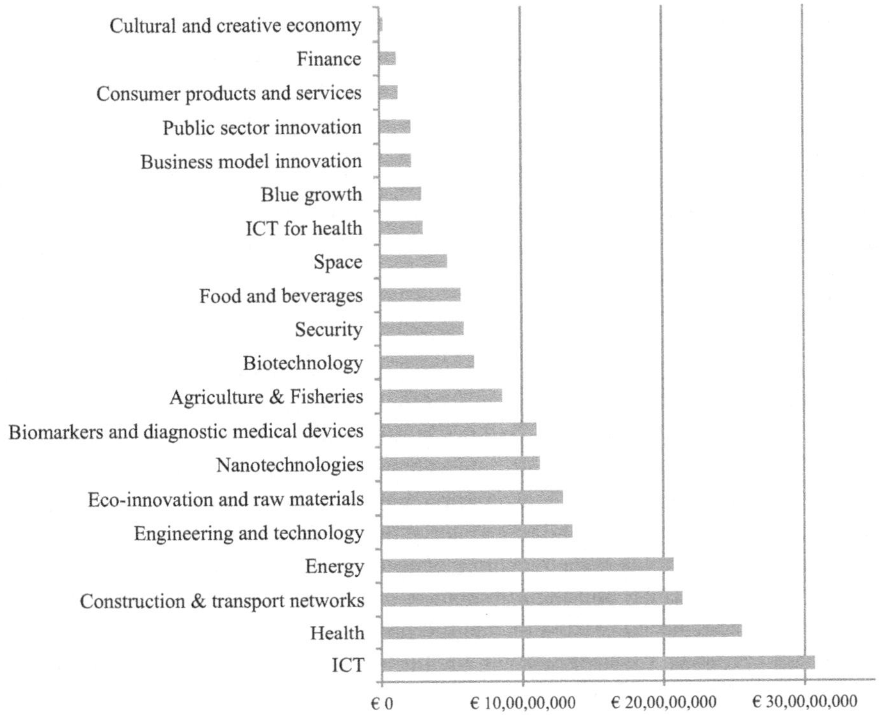

Fig. 4.18 Distribution by thematic areas of the Horizon programme funding (data processed from http://sme.easme-web.eu)

4.2.3 Horizon 2020 Projects on Waste Upcycling

The Horizon programme, with its significant budget of 78 billion euros, is succeeding in contributing to initiatives that concern numerous thematic areas. The most funded areas (Fig. 4.18) are those related to the following sectors: ICT (Information and Communications Technology), health, construction and transport networks and energy. More than 50% of the entire Horizon fund is allocated to these areas, as they represent areas that, if properly innovated, can produce cascade and wide-spread positive effects.

Among the projects relating to the construction sector, which involve the management, recycling or valorization of waste, it is possible to mention:

- The project "REACMIN [31]—Recycling asbestos containing materials into new eco-friendly secondary raw materials for further industrial processes"[46] deals with the activation of a treatment plant in which the thermally induced crystal-chemical

[46]REACMIN—Recycling asbestos containing materials into new eco-friendly secondary raw materials for further industrial processes (available at https://cordis.europa.eu/project/rcn/198894/factsheet/it).

transformation of 400,000 tons of Asbestos Containing Materials[47] (ACM) will be performed. The product deriving from this process (asbestos-free product suitable to be used as a secondary raw material) can be used for other industrial uses (concrete production, road building or environmental restoration).

- The project "Future Recycled Inert Concrete Made of Steelworks Residues" [34],[48] in line with the EU target (Directive 2008/98/EC) of 70% of C&DW to be recycled/recovered by 2020, proposes a strategy for replacing natural aggregates[49] (useful for the production of concrete) with the recycled steelwork waste residues.
- The project "Smart software platform for the economic and sustainable exploitation of dimension stone quarries" [35] [50] is promoted by the Tuscany Region and

[47]"*Before the discovery of its high toxicity, asbestos was for a long time considered as an extremely versatile and cheap raw material for building materials, automotive parts and even textile and other domestic products. In Europe, more than 40% of residential buildings were constructed before the 1960 s and in the 1990 s a large boom in construction still used asbestos. Health costs for diseases related to asbestos exposure amount over M€ 1.600 per year in Europe, so it is a main target to remove all remaining ACM. Only in Italy 2 billion m*^{2} *of asbestos-cement slates still cover roofs of industrial plants and civil buildings. Moreover, yet in 2012, rather than falling, worldwide asbestos production increased. The most extended current process for asbestos disposal is landfill in authorized sites. Associated costs are expensive, in the range of 150–300 €/Tn and no zero risk of fibres emission can be assured in the medium-long term. Other technologies based in thermal conversion exist but the only one at commercial level is up to 13 times more expensive than REACMIN system. The objectives of REACMIN project are: demonstrate the technology by building a pilot plant with optimal security measures and acceptable costs (<65 €/Tn). Test the resulting product suitability for the industrial production of concrete tiles Expand business model in other real industrial processes (glass, ceramics) Assure Zero risk of fibres release*" (from REACMIN description available at https://sme.easme-web.eu).

[48]FuRIC, Future Recycled Inert Concrete Made of Steelworks Residues (available at https://cordis.europa.eu/project/rcn/213323/factsheet/en).

[49]"*Natural aggregates are considered to be the best material used for the concrete production, in terms of mechanical performance (rigid skeleton structure) and cost efficiency of the final product. However, they are struggling to comply with the EU target (Directive 2008/98/EC) of 70% of C&DW to be recycled/recovered by 2020. They still are a natural resource, being directly deducted from nature, with collateral environmental and financial impacts. Hence, the concrete construction industry is in need of more sustainable and of reduced cost alternatives, willing to welcome at the same time an enhanced density product to be used in specific projects (harbor, seaside defense, etc.). Moreover, the steel industry needs cost-efficient solutions to treat its residues. FuRIC is a pioneering technology of green concrete, replacing the natural aggregate with recycled steelwork waste residues. In this way, we create a final concrete product made of up to 100% recycled material, maintaining the same structural performance, but with enhanced density (20% increase), at a lower final selling price (at least 50%). Also, our innovative product will contribute to decrease the environmental and financial impact of transporting and extracting natural aggregates (up to 40% reduction)*" (from FuRIC description available at https://sme.easme-web.eu).

[50]QUARRISMART, Smart software platform for the economic and sustainable exploitation of dimension stone quarries (available at https://cordis.europa.eu/project/rcn/211164/factsheet/it).

it deals with the issue related to the mining sector,[51] proposing an IT support that allows the sustainable management of the quarries. This is a very interesting project because it foresees a radical change in the way of imagining the quarry and its management, in this case nothing is left to chance or to the experience of the quarryman, but everything is weighted with precision instruments that can avoid or reduce waste production.

4.3 Results and Perspectives of the Conducted Experimentations

The study conducted, in line with what emerges from the reports on the progress of the funding programs [32], has highlighted that all the projects financed, within the European Union with public funding, contribute to the achievement of the EU objectives relating to the environment, climate and development. According to different methodologies and responding to different environmental emergencies, issues of all kinds are addressed, managing to cover a multiplicity of sectors and heterogeneous realities.

The in-depth analysis of the projects, in relation to the field of "waste", and more specifically with regard to the construction sector, has shown that nowadays there is a marked strongly interested in the topic of "waste", which is no longer seen as an element to be discarded but as a source of wealth and as an opportunity for lengthening the value chain. The initiatives are activated both in small and large production companies, this aspect varies according to the countries but it highlights that all the possible stakeholders are able to access these forms of financing. It should be noted that, in most projects, partnerships are built between different companies or interested parties, which makes it possible to achieve a cross-sectoral and multi-disciplinary nature of the project. Compared to the specific use of waste/scrap, different trends

[51]"*With a steady growth and more than 140 million tons of material traded yearly, the dimension stone sector represents today a dynamic industry, with the opening and development of new quarries in many countries, from Europe to the BRICS countries and others in Africa and the Far East. This growth of dimension stone quarries has raised attention to the natural resources and land degradation of mining activities often wild, non-economic and poorly managed. A solution to this problem has been developed by the developing of QUARRYSMART: a reliable, practical and easy-to-use software platform to support dimension stone companies in the pre-evaluation of quarrying costs and to assist them in turning their operations into sustainable, manageable and profitable extractive units, hence attracting investments while minimizing impacts on social, ecological and environmental systems*" (from QUARRISMART description available at https://sme.easme-web.eu).

emerge and some of these proceed towards up-cycling[52] (more valuable experiences) and many towards downcycling[53] [33]. It is clear that up-cycling is preferable to down-cycling, but in many circumstances due to the nature of the matter being treated this becomes inevitable. Moreover, it is possible to observe a strong interest in reviewing production processes to reduce the amount of waste and make it more compatible with new uses. The expectations on these initiatives concern their complete repeatability and the integration of competences to achieve positive effects on a large scale. However, it is clear that if successful initiatives are not disseminated and replicated, the investment made by the EU could become vain; therefore, for this purpose, it would be advisable to create a network in which the traceability of initiatives and best practices can be filtered through different reading keys linked to specific production sectors (see Chap. 5).

References

1. EEC (1992) Council Regulation No 1973/92 of 21 May 1992 establishing a financial instrument for the environment (LIFE), Brussels
2. EC (2000) Regulation (EC) No 1655/2000 of the European Parliament and of the Council of 17 July 2000 concerning the financial instrument for the environment (LIFE), Brussels
3. EC (2004) Regulation (EC) No 1628/2000 of the European Parliament and of the Council of 15 September 2004 amending Regulation (EC) No 1655/2000 concerning the financial instrument for the environment (LIFE), Brussels
4. EC (2007) Regulation (EC) No 614/2007 of the European Parliament and of the Council of 23 May 2007 concerning the financial instrument for the environment (LIFE+), Brussels
5. EC (2018) COM(2018) 385 final proposal for Regulation of the European Parliament and of the Council establishment of a Programme for the Environment and Climate Action (Life) and repealing Regulation (EC) No 1293/2013, Brussels
6. EEC (1996) Council Regulation No 1404/96 of 15 July 1996 amending Regulation (ECC) No 1973/92 establishing a financial instrument for the environment (LIFE), Brussels
7. EC (2014) Environment: €239 million available in 2014 for project proposals, press release, Brussels, 25 June 2014. http://europa.eu/rapid/press-release_IP-14-722_en.htm
8. EC (2018) EU budget: Commission proposes to increase funding to support the environment and climate action, press release, Brussels, 1 June 2018. http://europa.eu/rapid/press-release_IP-18-4002_en.htm

---

[52]The term upcycling was coined in the late 1990 s by the Belgian Gunter Pauli and the German Johannes Hartkemeyer (and later developed by McDonough and Braungart in the early 2000s) as opposed to the downcycling one, i.e. the passage that in the recycling process indicates the reduction of the material of a product in a basic form, often of a lower quality than the starting one.

[53]"*As we have noted, most recycling is actually downcycling; it reduces the quality of a material over time. When plastics other than those found in soda and water bottles are recycled, they are mixed with different plastics to produce a hybrid of lower quality, which is then molded into something amorphous and cheap, such as a park bench or a speed bump... Aluminum is another valuable but constantly down-cycled material. The typical soda can consists of two kinds of aluminum: the walls are composed of aluminum, manganese alloy with some magnesium, plus coatings and paint, while the harder top is aluminum magnesium alloy. In conventional recycling these materials are melted together, resulting in a weaker—and less useful—product*" [37].

9. EU (2013) Regulation (EU) No 1293/2013 of the European Parliament and of the Council of 11 December 2013 on the establishment of a Programme for the Environment and Climate Action (Life) and repealing Regulation (EC) No 614/2007, Brussels
10. EU (2006) Decision No 1639/2006/EC of the European Parliament and of the Council of 24 October 2006 establishing a Competitiveness and Innovation Framework Programme (2007 to 2013), Brussels
11. EACI (2013) First overview of the Call 2013 CIP Eco-Innovation. First application and market replications project, Brussels, 2013. Available at http://ec.europa.eu/environment/eco-innovation/files/docs/publi/first_overview_of_the_call_2013_cip_eco-innovation.pdf
12. CIP (2006) MEMO/06/226 CIP—innovation and entrepreneurship, ICT and intelligent energy, Brussels, 2006. Available at http://europa.eu/rapid/press-release_MEMO-06-226_en.htm?locale=fr
13. EC (2013) COM(2013) 2, Report from the commission to the European Parliament, the Council, the European economic and social committee and the Committee of the regions Evaluations of the Competitiveness and Innovation Framework Programme, Brussels
14. Bonifazi A, Giannetti A (2014) Finanziare l'impresa con i fondi europei, IPSOA
15. EC (2011) COM(2011) 834, Research, innovation and competitiveness package. proposal for a regulation of the European parliament and of the council establishing a programme for the competitiveness of enterprises and small and medium-sized enterprises (2014–2020), Brussels
16. EC (2011) COM(2011) 808, Communication from the commission to the European Parliament, the council, the European economic and social Committee and the committee of the regions. Horizon 2020—The Framework Programme for Research and Innovation, Brussels
17. LIFE08/ENV/E/126 (2013) Eco-stone. Sustainable system implementation for natural stone production and use, Final Report
18. LIFE+12/ENV/IT/000436 (2015) LIFE sustainable mission—test 1.0 of chemical industry for global sustainable organization as industrial total symbiosis and low energy and water, Project Technical Report
19. LIFE+12/ENV/IT/000374 (2017) Life plastic killer—innovative plastic pollutants removal for efficient recycled wood panels production, Layman Report
20. LIFE05/ENV/DK/000158 (2008) RECYCLING/SYMBIOSIS—Waste and Sewage Recycling and Symbiosis in Stone Wool Production, Layman Report
21. LIFE08/ENV/E/000167 (2013) ICEJET—Ice jet environmental technology pilot plant for drastically reducing waste produced by abrasive water jet cutting techniques, Final Report: technical version
22. LIFE+12/ENV/IT/000904 (2016) Life Frelp—full recovery end-of-life photovoltaic Lyman report
23. LIFE+09/ENV/IT/000188 (2015) Eco-cluster—environmental cooperation model for cluster, Layman's Report
24. LIFE+10/ENV/IT/000356 (2016) Gy.Eco—Gyproc Eco-friendly, Report
25. LIFE+08/INF/IT/000312 (2013) Promise—product main impacts Sustainability through eco-communication, Technical Report
26. ECO/13/630249, Wool4build. Improved isolation material for eco-building based on natural wool, information available at http://wool4build.com
27. ECO/13/630185, Insul-eco. Eco-innovative insulating thermal and acoustic panels made with recycled textile fibres, information available at https://ec.europa.eu/environment/eco-innovation/projects/en/projects/insul-eco
28. ECO/10/277233 (2014) ECOPLASBRICK. Innovative recycled plastic based panels for building field, Layman Report
29. ECO/08/239071, NATSTOCER. Sludge free-process for the production of innovative natural stone-like obtained by micro-structuring of sintered tiles, information available at https://ec.europa.eu/environment/eco-innovation/projects/en/projects/natstocer
30. ECO/13/630426, WINCER. Waste synergy in the production of INnovative CERamic tiles, information available at https://ec.europa.eu/environment/eco-innovation/projects/en/projects/wincer

31. REACMIN—recycling asbestos containing materials into new eco-friendly secondary raw materials for further industrial processes, information available at https://cordis.europa.eu/project/rcn/198894/factsheet/it
32. EC (2017) SWD(2017) 220 final Commission staff working document in-depth interim evaluation of Horizon 2020, Brussels
33. McDonough W, Braungart M (2002) Cradle to cradle: remaking the way we make things, FSG
34. FuRIC, Future Recycled Inert Concrete Made of Steelworks Residues. Information available at https://cordis.europa.eu/project/rcn/213323/factsheet/en
35. QUARRISMART, Smart software platform for the economic and sustainable exploitation of dimension stone quarries, information available at https://cordis.europa.eu/project/rcn/211164/factsheet/it

Chapter 5
A Virtual Marketplace for Waste Valorization

Abstract The reuse and recovery of waste are often slowed down and hindered due to the scarcity of information on its nature, but also to the impossibility of intercepting it before it is disposed of in landfill. Using a virtual marketplace through which handle waste streams could represent an interesting strategy for their systematic recovery. The chapter describes the results of a research carried out with the contribution of *Fondazione Fratelli Confalonieri di Milano* on the subject of the cross-sectoral valorization of waste in building sector. The chapter highlights the role and the importance that information and its standardization can have in a circular economy scenario aimed at the enhancement of pre-consumer waste. The chapter deals with: the dynamics that can manage a virtual marketplace aimed at enhancing waste recycle (systems for identifying potential users, successful experiences already conducted, etc.) and the databases that can be connected to the virtual marketplace itself in order to expand the possibilities for waste recovery.

Keywords Waste valorization · Cross-sectoral reuse · Secondary raw material

5.1 Usefulness of Information Standardization in the Valorization of Waste

The valorization of scraps/waste as a second raw material is an operation that can be activated constantly and efficiently only when a real and continuous demand comes from the market. However, the demand can only be activated when there is a valuable offer of materials that can be recovered periodically and with specific characteristics and standardizations. It is common for scraps/waste to be disposed of in landfills for reasons such as: lack of knowledge of the material to be disposed of; lack of information about how—and through which sector—it is possible to enhance the material. These motivations can no longer be a justification to continue to adopt a linear consumption model and it is necessary to find solutions that involve a radical

This chapter is authored by Marco Migliore, Politecnico di Milano, Department of Architecture, Built Environment and Construction Engineering, Milan, Italy. e-mail: marco.migliore@polimi.it.

M. Migliore et al., *Strategies for Circular Economy and Cross-sectoral Exchanges for Sustainable Building Products*, Springer Tracts in Civil Engineering,
https://doi.org/10.1007/978-3-030-30318-1_5

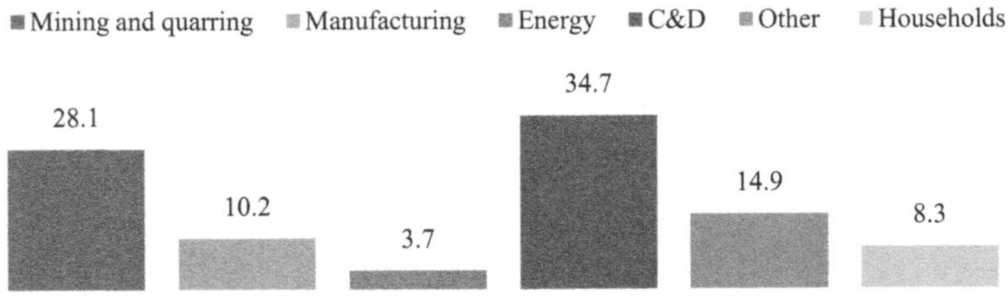

Fig. 5.1 Waste production in Europe for the year 2016, with percentage distribution in the main product sectors (Eurostat data)

rethinking of production and consumption cycles, shifting from the linearity of the classical system to a fully circular systems.[1] Furthermore the waste issue, besides being a problem from the environmental and sustainability point of view, is a topic of considerable importance also from the quantitative point of view, if we refer to the EU, every year we produce about 2.500 million tons on average of waste [1] for an average of about 4,900 kg of waste per inhabitant. This is a very high quantity that will rise if proper measures aimed at reversing the production trends are not applied. As regards percentage, the sectors that produce the greatest quantity of waste are those linked to extraction, which alone represent 28.1% of the total, and those linked to the construction sector, which represent 34.7% (Fig. 5.1).

Focusing on the Italian context and special waste, distinguishing between hazardous and non-hazardous waste, it should be noted that the total amount of special waste (hazards and non-hazardous) produced in 2016 (Fig. 5.2) grew by 3 million of tons and it is equal to 135 million tons [2] of which a percentage of 40.6% is attributed to the construction and demolition sector for a quantity equal to 54.8 million tons.

[1]In September 2015 the "Sustainable Development Summit" of the United Nations was held in New York, an epochal event that had the purpose of encouraging the adoption of common strategies among all states for sustainable development. During the summit, the "Transforming our world: the 2030 Agenda for Sustainable Development" was signed, which is the document of commitments on sustainable development to be implemented by 2030. Among these commitments, 17 global objectives have been identified (SDGs—Sustainable Development Goals) and 169 targets. The document is the result of a long preparatory process, which started during the "United Nations Conference on Sustainable Development, Rio + 20" and which was part of the debate on the future of the "Millennium Development Goals—MDGs", whose term was set for 2015. The objectives are universal in nature and they address all the countries. The objectives are based on the integration of the three dimensions of sustainable development (environmental, social and economic) as a system to eliminate poverty in all its forms. The new Agenda 2030 recognizes the close link between human well-being and the health of natural systems, therefore it directs every form of development towards forms of full sustainability. Among the Agenda goals there is the one linked to the "Responsible consumption and production" which includes among its targets: sustainable management and efficient use of natural resources; reduction of waste generation through prevention, reduction, recycling and reuse; adoption of sustainable practices and the integration of sustainability information into their reporting cycle. This topic is characterized by a double dimension: on one hand, action must be taken to manage resources more efficiently upstream of the system, by increasing productivity in both production and consumption processes; on the other hand, downstream it is necessary to avoid that everything that can still have utility and value is not disposed of in landfills, but that it is recovered and reintegrated into the economic system.

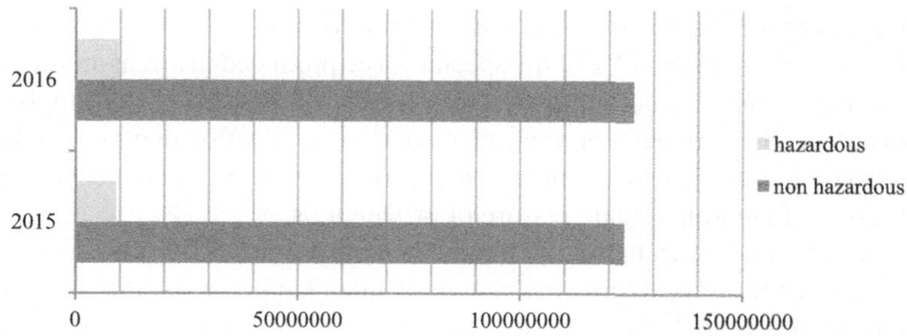

Fig. 5.2 Production of hazardous and non-hazardous special waste, reference period 2015–2016 (ISPRA data 2018)

With regard to the share of non-hazardous special waste produced [2], it should be noted that this is equal to 125 million tons (almost 93% of the total amount of special waste annually produced). The greatest production of non-hazardous special waste comes from the construction and demolition sector with a percentage equal to 43.4% of the total product, corresponding to almost 54.4 million tons (Fig. 5.3). This is followed by waste treatment & rehabilitation activities (26.9%) and manufacturing (19.4%), corresponding in quantitative terms respectively to 33.7 million tons, including the quantities of waste deriving from the treatment of waste urban areas, and about 24.3 million tons. It is clear that intervening with radical environmental improvements on the construction sector could represent a driving force for circular economy, since the quantity of involved resources and the heterogeneity of the companies would allow cascade effects on the entire economic system. The data [2]

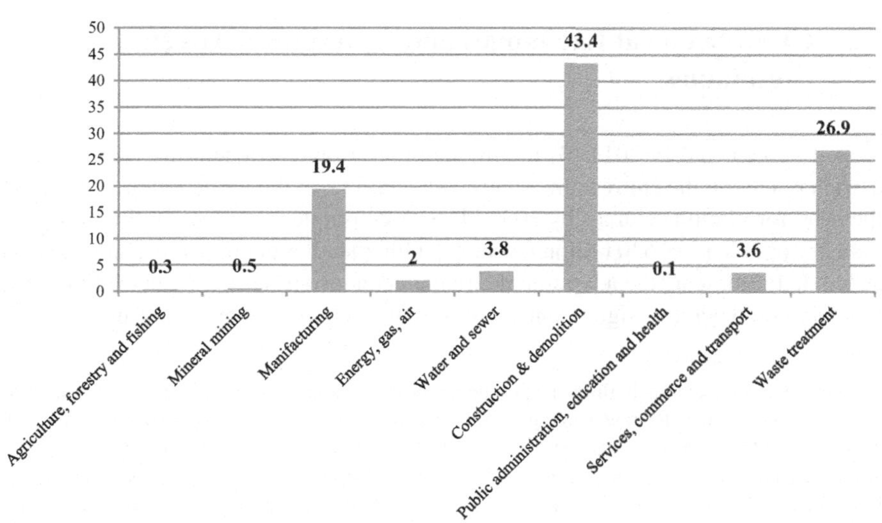

Fig. 5.3 Percentage distribution of total special waste production by business activity for 2016 (ISPRA data 2018)

presented underline that focusing even only on the construction sector would make it possible to reconsider over 40% of the special waste produced in a year and also there would be the advantages deriving from interventions upstream of the supply chain that can induce: the containment of virgin resources, a reduction in waste production (production efficiency), waste standardization (to prepare waste for recycling) and the reduction of environmental pressure of productions.

However, to activate scenarios of circular economy aimed at enhancing secondary resources, it is necessary to create a system of interchange of resources in order to be able to reach the goal of the "waste end". However, to achieve this goal it is necessary to convey the initiatives and simplify them, through the creation of organized networks that allow heterogeneous interested actors, spatially located in more or less vast territories to be able to communicate and be able to share resources, information, good practices, etc. Considering the typical Italian or European production context, this is characterized by more or less large companies that, after completing their production cycle, deliver scraps/waste to treatment or disposal centers without worrying about giving added value to their supply chain. This happens because companies do not have the possibility of reusing or enhancing their waste, a condition deriving from the lack of an adequate supply chain to be able to reintegrate and recover waste,[2] or from the impossibility of identifying who could be the potential users of their waste as a second raw material. It is clear that what is missing is a connection system that allows to share resources (scraps, waste, by-products, etc.), after having categorized and made them traceable. In the following paragraphs the outcomes of the research carried out with the contribution of Fondazione Fratelli Confalonieri di Milano will be introduced, proposing a platform for searching and sharing second raw materials.

5.2 Cross-Sectoral Exchanges and Synergies Between Companies

A study conducted in 2011 [3] highlighted that a more efficient use of resources, carried out along the entire value chain[3] [4], could favor the reduction of 17–24% of the need for virgin resources by 2030 [4] with consequent savings for the European industry of at least € 630 billion a year [5]. Other studies examined [6] have instead stated that, following the adoption of circular economy-based approaches, European industry could achieve significant savings on the cost of raw materials and raise the

[2]The reuse of scraps/waste in the same production process cannot always take place automatically, there are cases in which the raw material after the production process is transformed into its physical and chemical properties, and therefore it cannot be recovered without a pretreatment.

[3]The value chain is a model that allows the organizational structure of a process to be summarized, by defining a limited set of processes. Theorized by Michael Porter, the value chain is a useful tool for assessing, dynamically, if and when competitive advantage is achieved. The value chain disaggregates the activities strategically relevant for understanding the trend of costs and the main sources of differentiation. Source: Porter [4], The Competitive Advantage: Creating and Sustaining Superior Performance, NY: Free Press.

European Union's Gross Domestic Product to peaks of 3.9% [6], mainly through the creation of new markets and new products. All this highlights that the activation of synergies between companies can be rewarding for obtaining advantages and benefits on a large scale and, among the synergies that can be activated, there is undoubtedly the synergy relating to the recovery and enhancement of scraps/waste deriving from the processes productive, always if it is not possible to intervene upstream to avoid them and make the process 100% performing. The research conducted, also based on the contribution of other studies on the subject [7], examined and categorized the types of recovery that can be implemented. The purpose is to know how exchanges can take place and what are the factors that can influence their effectiveness. Taking into account some factors—such as the territorial scale of application, the number of companies that can be involved, the types of companies—it was possible to build a taxonomy of possible exchanges, highlighting related limits and incentives. An interesting contribution to the construction of this taxonomy takes into account the studies [7] conducted by Marian Chertow, Director of the Industrial Environmental Management Program at the Yale School of Forestry and Environmental Studies. Chertow has developed a taxonomy to describe the most common forms of material exchange by adopting a logic of industrial symbiosis and of recovery and valorization of production scraps/waste. The aim of his study [7] was to highlight the spatial implications and organizational logics, highlighting how these two factors are of fundamental importance to ensure the activation of sustainable production synergies. From the Chertow study [7] five types of possible exchanges emerge. In relation to distance, the exchanges that can be defined as more advantageous both from the planning point of view and from the economic/financial point of view, are those that can be implemented at the local scale (economic and environmental cost of reduced transport, organizational ease, etc.). However, it must be considered that, for some specific cases, even large distances can be overcome without economic losses and guaranteeing a fair environmental balance (avoided impact/induced impact).

The typologies identified by Chertow are:

- Through waste exchanges. This is a real exchange of waste between companies, which can be made more efficient if the companies equip themselves with an abacus of the materials they use for production and that derive from production. Chertow proposes to make this abacus available online which, if included in an organized database, can make exchanges more efficient. This type of exchange has a very wide range of action, obviously in proportion to the value generated by the recovery of a certain waste.
- Within a facility, firm or organization. This type of exchange is successfully implemented within the same company, within companies coordinated by a single management system or within companies that correspond to the different production phases of a given product or material. An exchange system of this type turns out to be very efficient at the local scale, however—if well calibrated—it can work even within a certain radius of action beyond which the environmental benefit is zeroed. Referring to the topic of waste, an exchange system of this type provides for an

immediate exchange of waste, since this is directly recovered and re-inserted into the supply chain without further treatment.
- Among firms co-located in a defined EIP. This type of exchange is effective when there is a common management system between companies that deal with the coordination of activities and resources. In this context consortiums or management platforms that, through joint actions, are able to make the system work smoothly are to be considered particularly suitable. The most appropriate scale of application is the local one, that of the district or industrial park, however the activity of the consortium can also extend "beyond the fence" involving other partners to complete the disposal of waste and/or convoy of external pre-consumer waste.
- Among local firms that are not co-located. This type of exchange is similar to that of the previous typology, however it refers to companies not placed near each other. In this case, the exchanges are designed to overcome considerable distances without too many problems, however they are always organized systems that feed themselves and, also in this case, there is a common management system that imposes directives and makes the passages fluid. In these cases, it is possible to speak of an Industrial Council, since these are operations that can occur even beyond the regional level.
- Among firms organized "virtually" across a broader region. Given the high cost of travel and other critical variables that fall within the decisions regarding the location of companies, there are very few companies willing to relocate exclusively to engage in an industrial symbiosis system. This exchange system involves a virtual rather than a geographical connection. Although the companies involved in a system of this type are far from each other, the possibility of a wider exchange—perhaps of a regional type—widens the potential for re-use of by-products since the range of companies available to participate is varied and expanded. An important feature that this type of exchange can have is the possibility of being able to include in the system small peripheral companies that actively contribute to feed the processes.

The scaling factor is of primary importance and cannot be neglected when connections between companies are established. An underestimation of this factor could invalidate the project strategy and it could affect the expected economic/environmental result. Together with the scalar factor, also the territorial factor is fundamental to obtain advantages from the cooperation. Trigilia [8], an expert in economic sociology, addressed the issue of the importance of the territory for the understanding and analysis of industrial districts.

The consideration of the territorial factor [8] has allowed us to estimate the productive force of the industrialized territorial areas, managing to explain, even ex post, the reasons why in some areas we can find a high index of industrial development and in others not. Trigilia, questioning the nature of local development, argues that territorial policies have a full centrality in defining competitive geo-economic scenarios. He also attributes them the power to favor or not innovation processes—creating a rift with the established local historical and geographical conditions—which can be considered useful only in the phase of recognition of territorial vocations. The productivity of the territories [9] is therefore correlated to the intrinsic ability to create local collective goods[4] that can contribute to increase the competitiveness among the companies: this happens both because the general costs are lowered and because the capacity for innovation is favored by the cooperation. This is not a result deriving only from local traditions, but rather from national cooperation between local authorities, companies, institutions and research centers, which lead to favoring the formation of external economies, both material and immaterial. Therefore, referring to what reported, it emerges that the factors that can influence the possible exchanges/displacements of material are two: scalar factor and territorial factor. On one hand, the distance between the actors involved and, on the other, the interest of the actors involved in wanting to do better. Looking at the national and international scene, some peculiarities emerge which, even in this case, can be summarized in five types of exchange/recovery of scraps/waste material (Fig. 5.4).

Specifically, the second scale was considered considering the following scales: company, industrial district and extra-district. Furthermore, scaling has been taken into account by associating a certain relevance to these three types of organization (both companies and districts that the super-districts must act for the common interest of the companies or associations they represent). In Figs. 5.5, 5.6, 5.7, 5.8 and 5.9 the types of exchange are described and outlined.

As emerges from the study conducted and from the developed taxonomy, the predominant factors for a correct structuring of resource exchange synergies are: scaling and territoriality. These two factors, if well combined, are the basis for being able to implement circular economy scenarios in an industrial environment. However, in order to be active and proactive, these networks must be supported by other information that make it possible to trace companies with similar interests and/or companies that produce scraps/waste useful as a second raw material. For this purpose, it is required a coding and management system for waste that can be managed within a platform/marketplace.

5.3 Guideline for a Waste Virtual Marketplace

5.3.1 Identification and Possible Codification of Scraps

Knowledge of scraps/waste is essential to be able to transfer and/or re-enter it in other production contexts. One of the problems that currently limits these practices is the

[4] According to Achille Flora "local development, in its various forms, is accompanied by the capacity of strategy of public and private subjects, by their commitment to coordinate with formal and informal tools, to support a shared development design. Leadership and quality in the ruling class are not created by decree, but they can be stimulated by intelligent policies, which use incentives to mobilize and empower local society, even financially". Source: Flora, A., 2008, *Lo sviluppo economico: i fattori immateriali, nuove frontiere della ricerca*, Franco Angeli Editore, Milano, pp. 180–181.

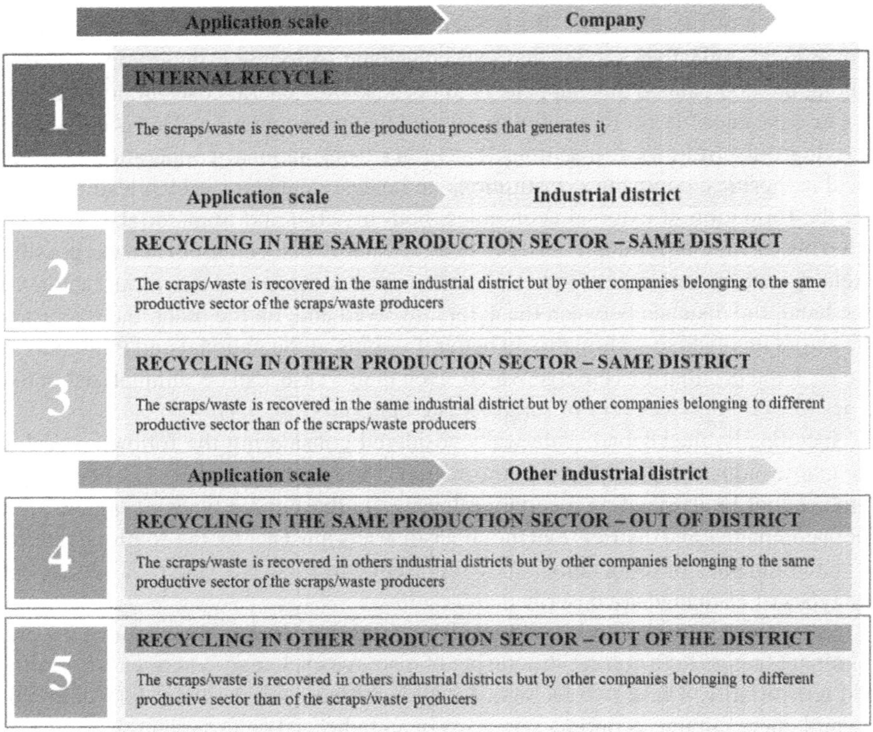

Fig. 5.4 Scheme of possible exchanges of scraps/waste between companies

lack of knowledge of what, how and how much is produced in terms of scraps/waste. Many waste materials are potentially reusable, but without a system that is able to trace and identify their position/quantity/quality it is difficult to be able to implement any type of recovery. A strategy that can be used to promote the re-use and enhancement of pre-consumer material waste can be the strategy that has as main objective an easier access to information by potential interested parties (end users).[5] The activation of industrial synergies based on recycling depends on the possibility of being able to trace recoverable waste, since the characteristics that distinguish them are not always compatible with the possible reuse [10] or re-insertion[6] in the supply

[5]In many situations scraps/waste is destined for landfill disposal due to difficulties that place companies in the face of bureaucratic impediments. For example, according to the art. 10 of Legislative Decree No. 210 of 2010 (currently in force in Italy), everything that cannot be directly recovered or transferred to another supply chain must be transferred to landfills. There are also quantitative and temporal constraints that require companies to have to undo what they do not need (at most 30 m^3 stored in the company of which at most 10 m^3 of hazardous waste, disposals at least annually if the quantity held does not exceed the above limit, or every quarter).

[6]Three types of reuse can be identified: primary, secondary and tertiary. The primary reuse is the most advantageous because it allows the waste to be used directly as a second raw material without further treatment (for example the raw waste in the processing of ceramics). Secondary reuse, on

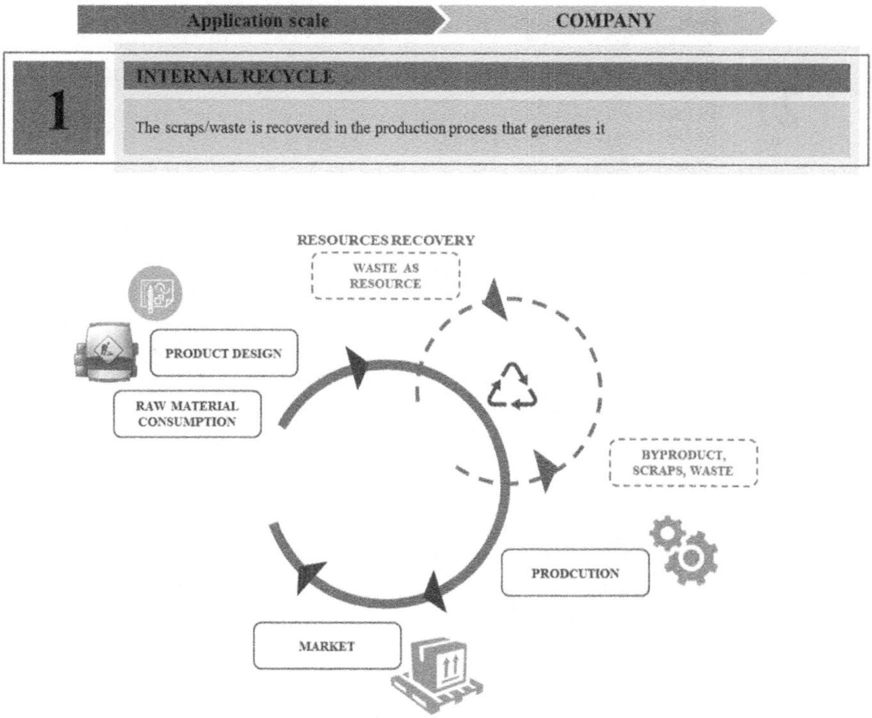

Fig. 5.5 Internal recycling involves the reconsideration of the scarps/waste directly in the source chain. This operation is not always possible, but many companies are working to recover second raw material by reprocessing it (see ceramic companies and how to recover waste from processing, cutting and cooking) or to characterize it in order to avoid contamination of the waste and make it suitable for recovery in the supply chain

chain of origin or in other supply chains. To this purpose, the presence of descriptive information on the characteristics of potentially exploitable[7] waste is fundamental but, above all, the information must be well organized and related to the various cognitive and evaluation purposes that may be necessary in order to implement continuous recycling plans. It should be remembered that waste is never considered in its

---

the other hand, requires at least one mechanical processing for its recovery (e.g. the cut-outs of the wooden panels that are crumbled and re-assembled) and it can foresee a decline in the quality of the original material. While the tertiary reuse is the most complex and it involves at least a functional chemical treatment to obtain a new product identical to the original one. Source: Pernice and Mininni [10], *Il sistema normativo e tecnico di gestione dei rifiuti*, Ipsoa editore.

[7]For example, the waste resulting from the processing of stone is potentially always recoverable, however, on the basis of the composition of the slab that is cut (the rock is classified into sedimentary, metamorphic and magmatic, and each of these rocks has different characteristics) it can be destined for different supply chains. For example, those richer in limestone can be used for the production of cement, those that contain more quartz can be used to produce artificial stones, etc.

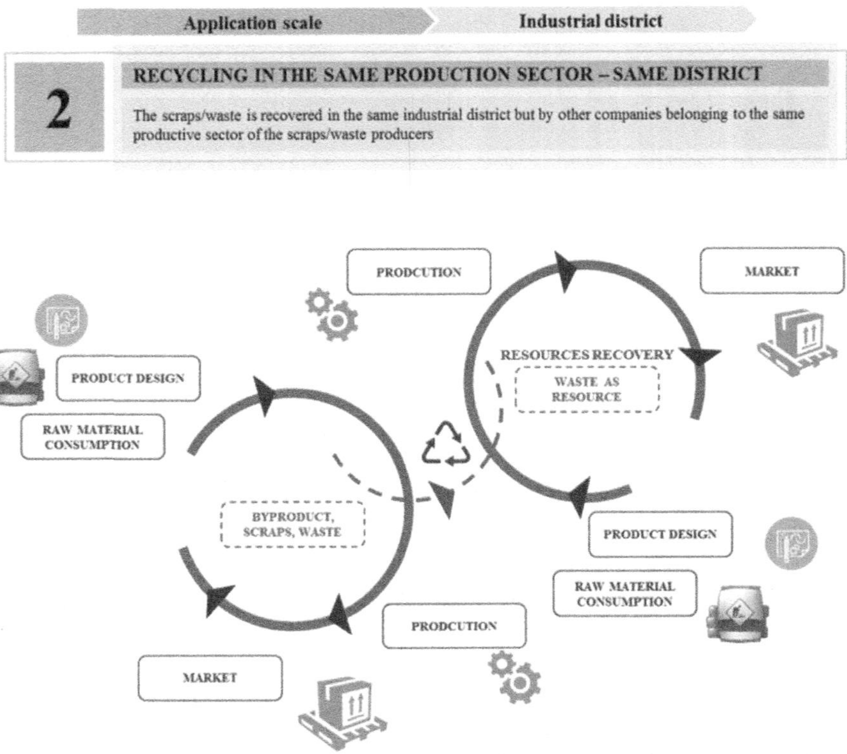

Fig. 5.6 In many cases recycling can take place in the same productive district (maintaining a reduced scaling in the transfer) but in different companies belonging to the same product area. A valid example of this recycling procedure is that of the lithic sector, where waste often becomes a raw material for companies connected to the lithic chain, such as for example oil mills that can produce aggregates for the production of recomposed stones

potential as a second raw material, due to insufficient information[8] associated with it (their spatial location, quantity, morphology, technical characteristics, availability over time, etc.).

[8]To make any scraps/waste recovery strategy workable, the nature and the availability of the waste must be clear over time. For example, in the steel production chain, they are generated by-products that can be reused in the mixture that constitutes the asphalt road surface, however it is known that, while the foundries never interrupt their production, the companies that stretch asphalt tend to work less in winter: this creates an imbalance in the exchange of second raw materials that forces steel mills to dispose in landfills the excess waste held in the company.

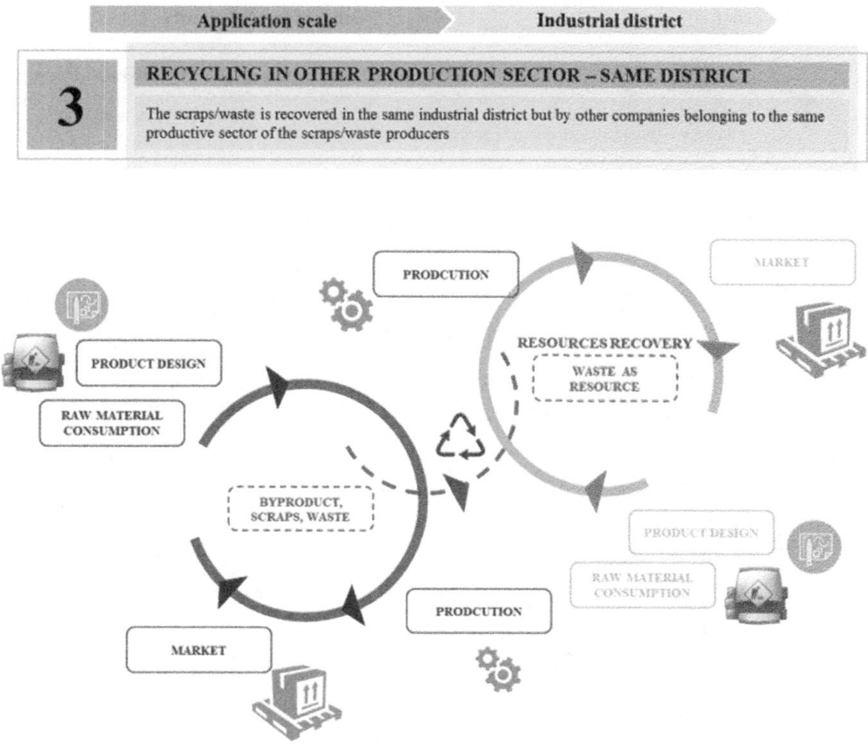

Fig. 5.7 In the same district different production realities can coexist and the processes of recycling—while maintaining reduced scalability—can be activated between companies attributable to different product sectors. See the case of the agricultural sector (present in many Italian manufacturing districts) which can supply reusable waste materials as a raw material in other sectors, such as construction (straw and hemp bricks)

Currently waste is identified with the so-called EWC codes,[9] a unified system at European level that allows to recognize the waste with respect to the sector and the production process that generates them, as well as for their possible dangerousness. However, in some cases, this code remains rather generic. In particular, for the purpose of an ordinary disposal the code is very clear, but for the purposes of

[9]EWC is the acronym of "European Waste Catalog". The Directive 75/442/EEC defines the term "waste" in the following way: "any substance or object that falls within the categories listed in Annex I and which the holder discards or has decided or is obliged to discard". Annex I is commonly known as the "European Waste Catalog" (EWC) and applies to all waste, whether destined for disposal or recovery. The European waste catalog is subject to periodic revision. The EWC codes are numerical sequences, composed of 6 digits combined in pairs (e.g. 03 01 01 bark and cork waste), aimed at identifying a waste based on the production process from which it originated. The first group identifies the chapter, while the second usually identifies the production process. The codes are inserted into the document "Notices from European Union Institutions, bodies, offices and Agencies. European Commission. Commission notice on technical guidance on the classification of waste (2018/C 124/01)".

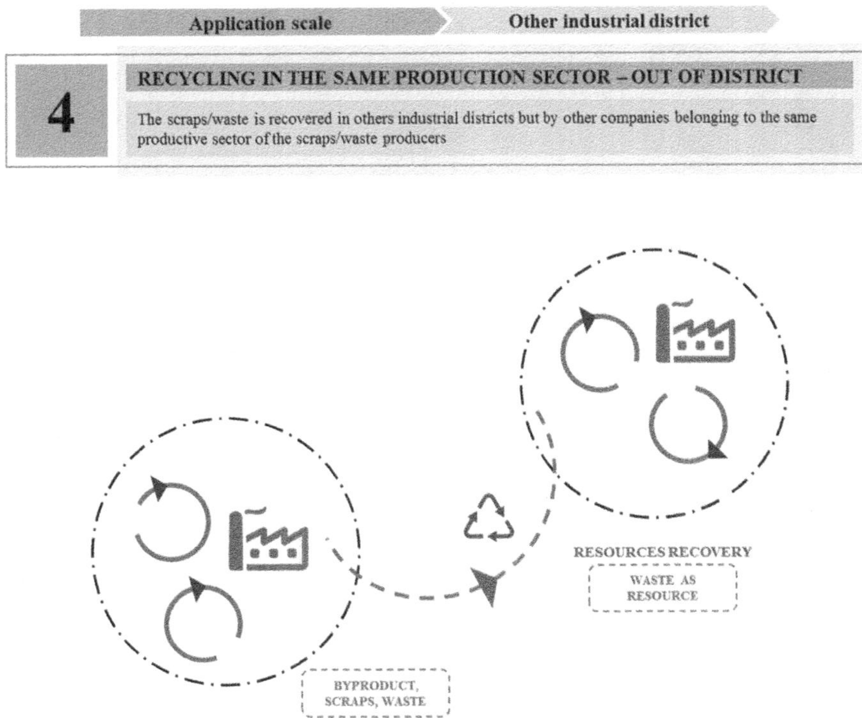

Fig. 5.8 An atypical circumstance, but still traceable in the Italian and foreign production reality, is that of transferring scraps/waste material to companies in the same sector but located in different industrial districts. This operation can be successfully activated when a balanced environmental and economic advantage is guaranteed

a possible recovery/valorization, the code appears to be cryptic and of little help. In fact, the EWC code was not created for this purpose and it is unable to transmit information relating to the quality and characteristics of the waste. However, it represents a good starting point in order to be able to construct a "speaking" code that uniquely identifies a waste by associating it with a series of useful information for the purpose described so far. The proposal of this research project intends to associate more information exploiting the relationship among the NACE code (Ateco 2007[10]), the EWC code and the Abaco code (Fig. 5.10). The proposed code is divided into 4 parts:

[10]Automatic coding tool that allows to assign an Ateco 2007 code based on a summary description of the economic activity provided by the user. This classification is the Italian version of the European nomenclature Nace Rev.2, published in the Official Journal on December 20, 2006 (Regulation (EC) n.1893/2006 of the EP and of the Council of 20/12/2006). The 2007 Ateco was defined and approved by a Management Committee. It involved the participation, in addition to Istat (which was the coordinator), of numerous institutional subjects: Ministries, Bodies that manage the main administrative sources on companies (tax and chamber of commerce, social security institutions, etc.) and main business associations.

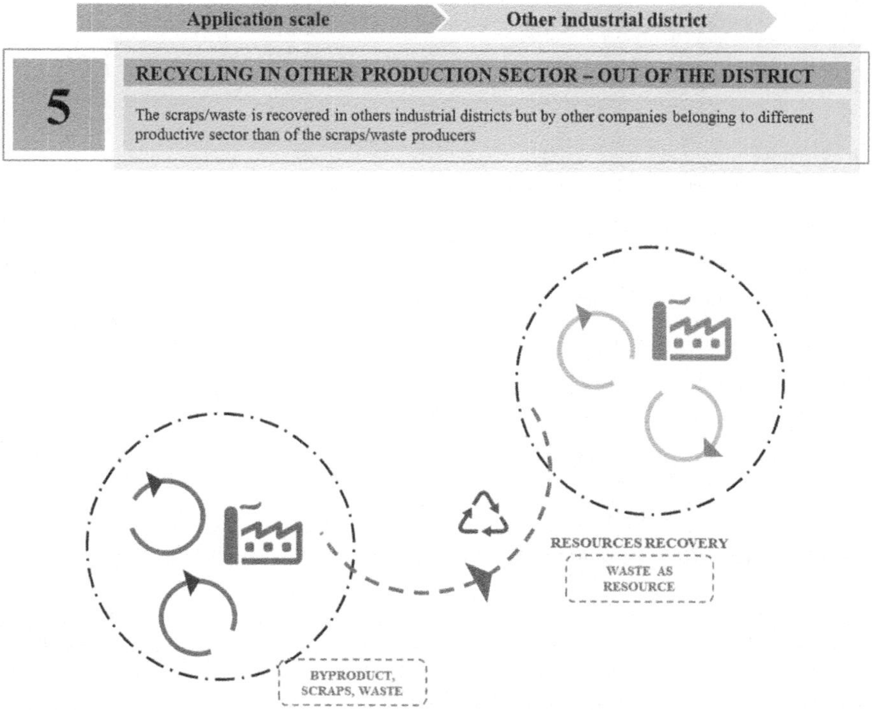

Fig. 5.9 This form of recycling can be considered as the most interesting and it opens up more possibilities for recovery. The waste produced by any company in any district can become a second raw material for other companies located in any other place. This is the objective of the research project, always ensuring compliance with environmental and economic parameters

- the first part groups under the main production chain all the constituent processes. It may not be codified and can only be referred to as a trace for specific research on the supply chain (Table 5.1);
- the second part of the code is taken from the NACE database (Ateco 2007) and it identifies the processes that make up the main supply chain;
- the third part identifies the waste/waste according to the EWC code;
- the fourth and last part identifies the refusal according to the Abaco code. The Abaco code is the object of the research and consists of 3 fields separated by points that identify three main peculiarities of the waste useful for its identification for recovery purposes (Fig. 5.11).

Figure 5.12 shows the trace followed to construct the Abaco code, which considers as useful for recycling purposes: the type (pre-scrap,[11] byproduct, sludge, powder, post-scrap[12]), the size (small, medium, big) and the purity of the waste (pure, mixed,

[11]It identifies pre-processing waste (typically not contaminated).

[12]It identifies post-processing waste (which may contain other substances).

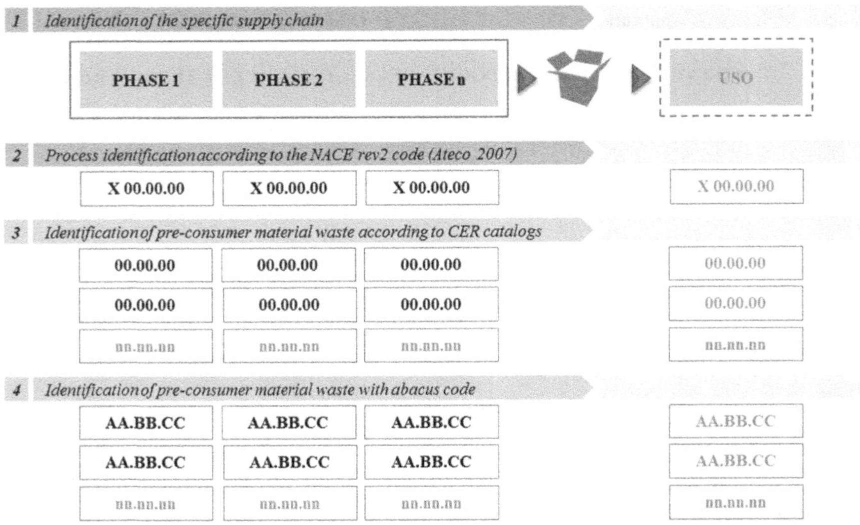

Fig. 5.10 Structure of the scraps/waste identification code

Table 5.1 Groupings by sector of NACE activities. Example conducted on the natural stone supply chain and on the supply chain of the wood for construction (based on "Notices from European Union Institutions, bodies, offices and Agencies. European Commission. Commission notice on technical guidance on the classification of waste (2018/C 124/01)")

Supply chain	Code	Material	Description
	[NACE rev.2]	[EWC]	
Natural stone	B 08.11.00	01.01.02	Waste from mineral non-metalliferous excavation
		17.05.04	Soil and stones other than those mentioned in 17 05 03
	C 23.70.10	01.04.10	Dusty and powdery waste other than those mentioned in 01 04 07
		01.04.13	Waste from stone cutting and sawing other than those mentioned in 01 04 07
		01.04.99	Waste not otherwise specified
	C 23.70.20	01.04.07*	Waste containing hazardous substances from physical and chemical processing of non-metalliferous minerals
		01.04.13	Waste from stone cutting and sawing other than those mentioned in 01 04 07
		01.04.99	Waste not otherwise specified

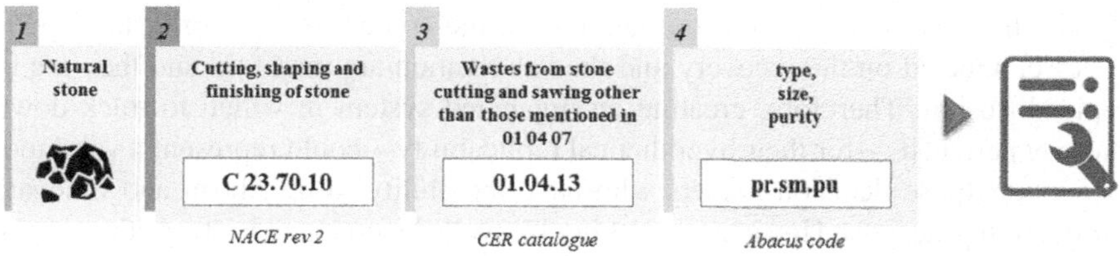

Fig. 5.11 Structure of the scraps/waste code

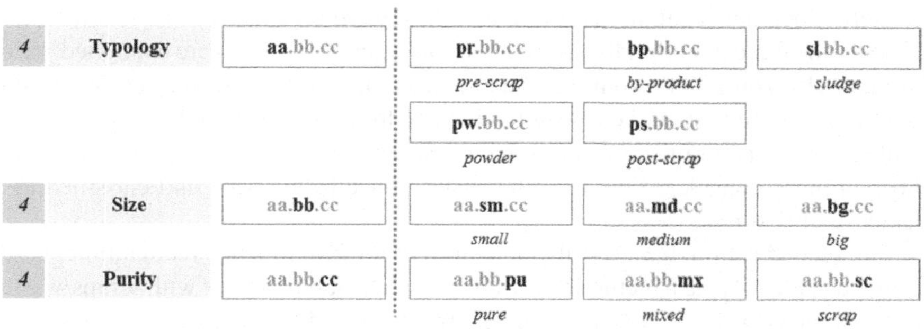

Fig. 5.12 Structure of the abacus code

scrap[13]), obviously—for completeness—a technical sheet[14] must be drawn up. This technical data sheet must contain mechanical, physical and chemical specifications.

5.3.2 Definition of the Relationships Between Scraps/Waste and Possible Recoveries

After defining the system to unambiguously code and identify scraps/waste, the aim of the research was to provide indications in order to be able to cross-reference the information relating to scraps/waste and their recovery. In some cases recovery and valorization can be immediate, in other cases a study and a preliminary evaluation are necessary to understand what, how and where it can be done. In order to achieve this objective, two actions were carried out: the identification of consolidated and replicable recycling and valorization scenarios and the definition of possible relationships—through a matrix—between the information gathered about the recovery and about the valorization of scraps/waste.

[13]Identifies the finished product clippings (dimensional adjustments).

[14]The scraps/waste technical data sheet, as for the main products of the supply chain, aims: to provide all the information needed by hypothetical users to understand their nature, and to assess whether the waste is adequate for the supply chains and production processes for which its integration was planned.

The first action starts from the fact that in the contemporary world the experiences conducted on the recovery and the valorization are manifold and they are in rapid diffusion. Therefore, creating an organized system in which to track down these experiences—for their hypothetical replicability—could represent a valid tool to support those decision makers who have the ability to transform and innovate production processes. However, searching for well-documented best practices is not easy. If the initiatives were conducted privately by private companies, the modus operandi is never disclosed (for obvious industrial policy reasons); instead, when these eco-innovation processes take place through public funding, there is a wealth of useful documents and information and, thus, good practices can be identified. In Chap. 3, the three main funding programs conducted in the EU were examined: Life, CIP and Horizon 2020. From this examination, the most interesting projects were identified with respect to the research topic and they were analyzed in depth. In particular, their structure and evolution have been carefully analyzed and, in addition, those projects considered the most interesting have been coded and classified in a database of best practices.

The second action, based on the outcomes of the best practice recognition phase, is aimed at identifying a method that allows to relate best practices with scraps/waste. The system used (Fig. 5.13) to relate best practices and scraps/waste involves some reading keys that act as filters in the search.

In particular, there are three types of reading keys:

- Type of innovation proposed: new production process, innovative production process, new product with recycled content, innovative product with recycled content, service.
- Type of activity on scraps/waste: reduction, elimination and recycling.
- Type of product sector and resources involved: EWC code, Nace Code, Omniclass 21, UNI 8290.

The first two keys of reading work on best practices at a general level, identifying strategies that can concern a specific sector, but which can also be indicative only of the procedure followed. The third key of interpretation becomes very specific and it allows to identify and select best practices that concern in a very direct way a specific sector or a specific scrap/waste. In particular:

- Nace code of the applicant. It allows to filter the research projects studied according to the Nace code of the company that requested the financing. This is a useful identification system because it allows companies belonging to the same product sector to identify the activities aimed at eco-innovation activated by other companies and therefore potentially replicable.
- Nace code of the receiver company. It allows to filter the research projects according to the Nace code of the project destination (the funding could have been requested by a company not involved in that type of activity or by a company that does not produce a given scrap/waste). This is a useful interception system because it allows companies belonging to the same product sector to identify which activities, aimed at eco-innovation and potentially replicable, have been activated by others.

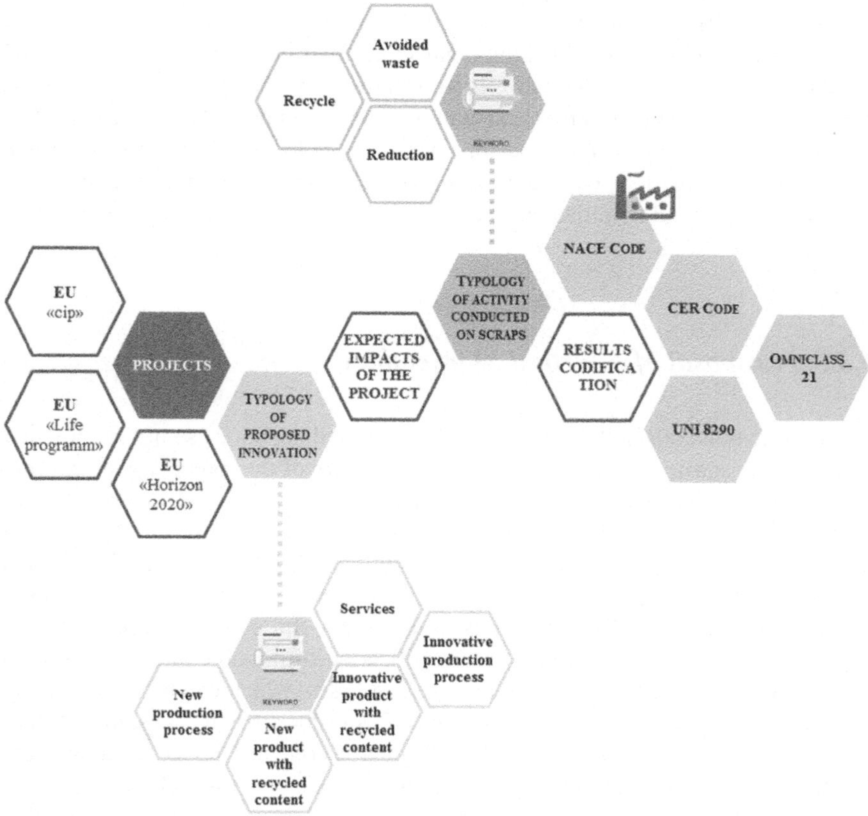

Fig. 5.13 Scheme of the analysis and classification system conducted on the Life, CIP and Horizon 2020 projects

- EWC code. It allows to filter the research projects surveyed in the best practice database according to the EWC code of the scrap/waste materials. This is a very useful identification system because it allows companies a wider possibility of identifying possible scenarios of recovery and environmental improvement.
- UNI 8290. It allows to filter the research projects surveyed in the best practice database according to the UNI 8290 standard. In the case of this research project—relating to the construction sector—the UNI 8290 standard makes it possible to trace the manufacturers of specific technological systems, who have experienced scenarios of recovery and enhancement of the second raw material.
- OMNICLASS_21. It allows to filter the research projects registered in the best practice database according to OMNICLASS_21. Specifically, it is possible to trace the manufacturers of specific technological systems who have experienced scenarios of recovery and enhancement of the second raw material.

The best practice database, structured according to the terms proposed, can become a valid tool for establishing connections and orienting industrial or territorial policies. This tool will aim to foster a starting knowledge base to start studies that can lead companies, districts and administrators to understand what can be done with respect to their production sectors. The logic is to foster a fluid and constant exchange of information between those who have already done (and have done well) and those who intend to initiate business transformations aimed at a greater sustainability.

5.3.3 Proposal of a Virtual Marketplace

The information collected through the taxonomy and the database of best practices represent the starting point for defining the structure of the inter-sectoral information platform,[15]whose goal is to put in communication offer and possible demand for pre-consumer waste, potentially reusable/recyclable, in order to activate virtuous circular economy mechanisms. Such an outlined platform can develop a virtual marketplace and it can adapt to multiple territorial scales, as well as to rapid changes in production realities. The purpose is to facilitate—through the access to knowledge by a multiplicity of actors—the identification of possible alternative uses of scraps/waste and the related transformative processes, which can be activated with a short supply chain, and with a defined local/supra-local basin. Thus, the platform represents the systematization of all the information collected (waste taxonomy, best practice database) and it is accessible to all the hypothetical users (Fig. 5.14).

The platform was called "Ri-scarto", as the research project funded by "Fondazione Fratelli Confalonieri", and its use method can be articulated according to the following four actions:

---

[15]Currently in the European community there are various platforms, which allow to follow the latest environmental innovation initiatives related to products and production systems, as well as various agencies supporting these initiatives. In this regard it is possible to mention:

- OREP—Online Resource Efficiency Platform. It is a platform in which contributions on the subject of resource efficiency are collected. Shared between all member states, the platform collects information, strategies, evaluation indicators, etc., which can help companies that want to move towards green and sustainable markets.
- EASME- Executive Agency for Small and Medium-sized Enterprises. It is a structure of the European Commission to manage different forms of community programs (COSME, EEN, HORIZON 2020, LIFE, FTI, etc.). Within this agency, it is possible to find information and studies of various kinds useful to promote innovation and support for entrepreneurial activities.
- EUR-ISA—European Industrial Symbiosis Association. It is a structure that seeks to accelerate the mechanisms of industrial symbiosis, creating synergies between companies that pursue the same goal. The aim is to favor the creation of networks on a local, regional and global scale to achieve a widespread industrial symbiosis.

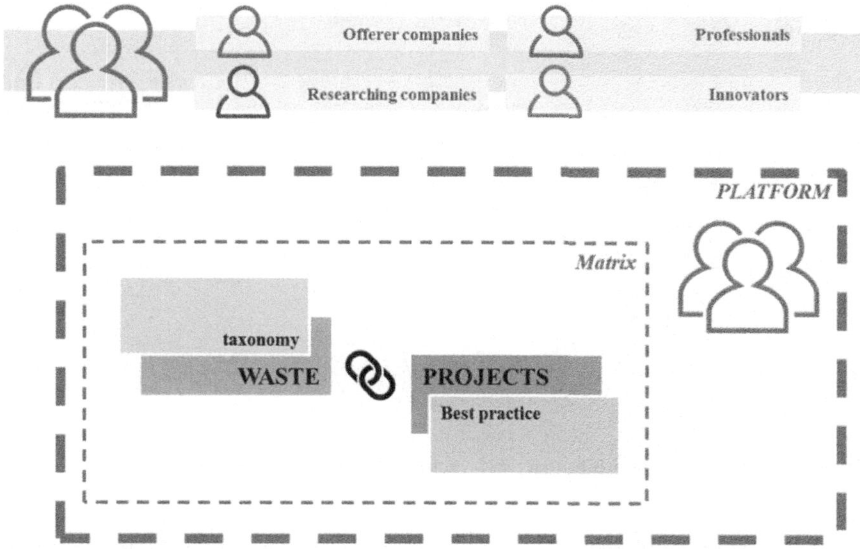

Fig. 5.14 Schema relazionale tra le banche dati e gli users della piattaforma

Action 1 Registration to "Ri-scarto" platform. The platform is accessible through the registering/profiling of the company/subjects interested in the proposed recycling and valorization scenarios. The platform manager has not been formally defined yet, as it may vary according to the scale of application (industrial district, province, region, country, etc.) and/or the product sector (e.g. the platform could be restricted and managed by specific associations of category and it could only work for a specific sector). Clearly the platform design was conceived to be extended as much as possible, since limiting it in size could reduce the chances of recovery. Once the credential has been obtained from the manager, the user will have to profile (action 2) to be able to carry out his research aimed at activating industrial synergies linked to sustainability.

Action 2 User profiling. The platform has been designed to be used by different users with different objectives. The expectations deriving from the use of the platform are those of favoring the exchange of scraps/waste through the technical profiling of users. This operation becomes possible depending on the connections established between the needs of the different users. Among the users there are: scraps/waste producers interested in offering them; companies or individuals interested in acquiring second raw material, professionals interested in acquiring materials with recycled content; stakeholders interested in initiating innovative recovery processes within production chains (Fig. 5.15).

The profiling involves the definition of the scope of operation of the company/user through the definition of the NACE code and of the specific

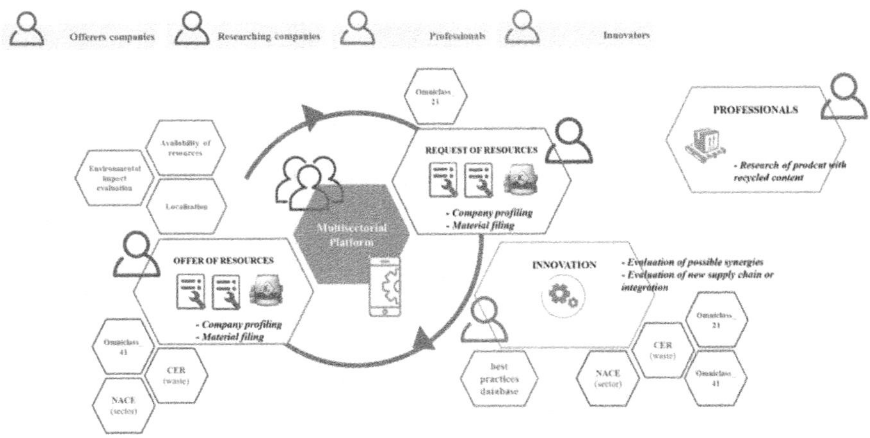

Fig. 5.15 System of relationships that can be activated between different users

sector of reference. This first information allows to filter and identify companies that belong to the same sector and that, since they have common interests, can establish industrial connections of different nature. Subsequently, the information required to complete the profiling covers 4 macro areas that will allow to trace the resources (primary, secondary) and the products sent-off from the production processes. The macro-areas identified are raw materials (which are used to produce the main product of the supply chain), products, by-products (if present) and waste (Fig. 5.16). Each user can fill-in this information according to what he is offering and what he is looking for. A greater quantity of information increases the probability of being able to establish connections that comply with the principles of the circular economy.

Action 3 Search for possible industrial partners from which to supply secondary resources and to whom to allocate them. The platform implemented with all the information described up to now, associated with the database of best practices and with the taxonomy of waste, makes it possible to activate the search for industrial partners with which to establish exchanges of second raw material. The main types of exchange that can be activated are two: the search for sectors to which to allocate[16] (Fig. 5.17) the second

[16]Any producer of scraps/waste and/or by-products, aware of the value inherent in this material, aims to identify possible partners to whom the material can be transferred. The possibilities that will lead him to identify the best industrial partner can be: search for raw materials similar to his own waste; perform a search in the database of best practices of the initiatives that have enhanced the scrap, and then proceed with the enhancement or with the search for geographically close companies with which to establish synergies. The last step before starting the actual exchange is the territorial evaluation (e.g. where the waste material is located and where it must be transported, how much its recovery will cost in environmental and economic terms).

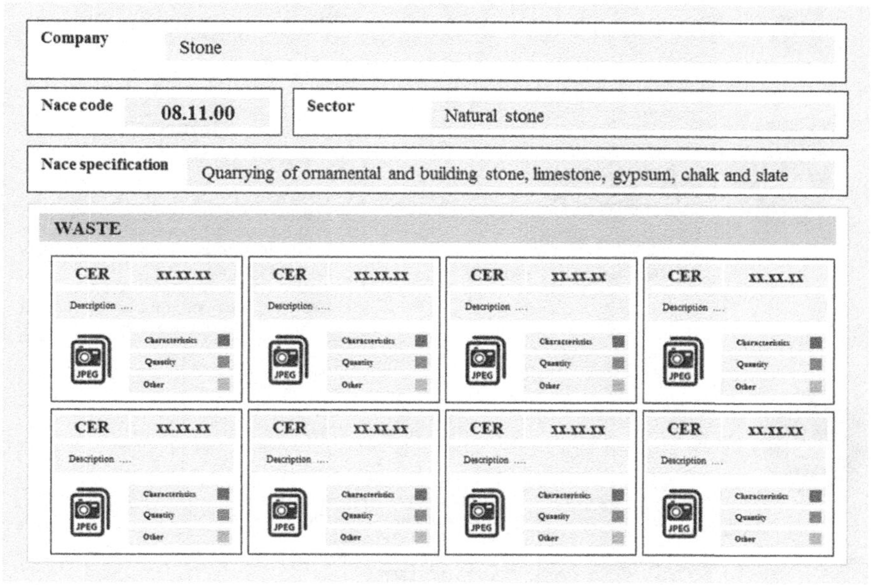

Fig. 5.16 Scheme of the data collection system for the waste section. In this way, each company can profile the type of scraps/waste produced and make it available to other hypothetical users. This operation is carried out for raw materials and for products, as well as for by-products and for waste

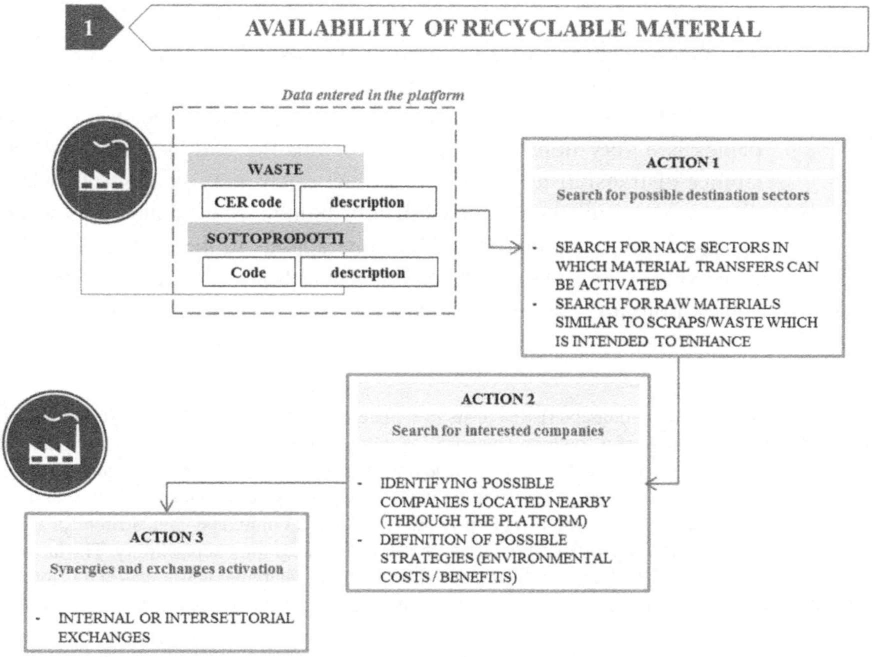

Fig. 5.17 Outline of the procedure for identifying possible industrial partners

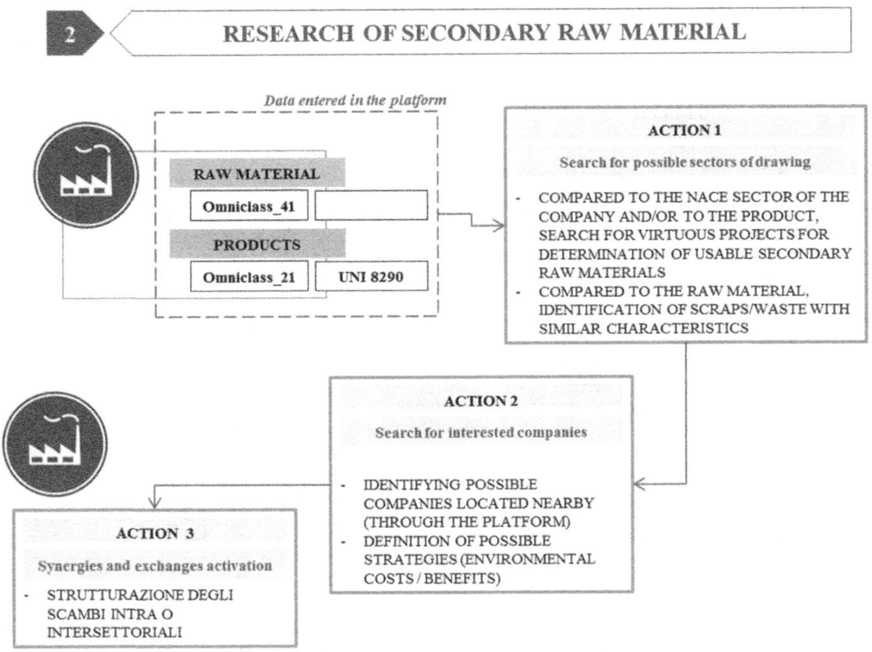

Fig. 5.18 Outline of the procedure for identifying possible industrial partners

raw material and the search for sectors from which to supply[17] (Fig. 5.18) the second raw material.

Action 4 Geographical location of resources. The geographical location of companies is a very important factor for estimating and assessing the feasibility of transferring resources. In fact, there may be cases in which distance does not significantly affect economic aspects and cases where distance becomes a problem. The same applies to the evaluation of the avoided impacts and the induced impacts linked to the transfer. The possible actions are, therefore, the quantification[18] of the available resources

[17]Any company, given the characteristics of its raw material, can look for waste/waste with characteristics similar to those of its own raw material. In this way industrial synergies can be activated between companies. The possibilities that will lead him to identify the best industrial partner can be: search in the database of best practices for initiatives carried out by companies that have enhanced scraps and waste with respect to their main product (UNI 8290, Omniclass_21); search for waste similar to its raw material; identify companies with respect to geographical proximity. The last step before starting the actual exchange is the territorial evaluation (where the waste material is located and where it must be transported, how much its recovery will cost in environmental and economic terms).

[18]Through the platform, companies declare the amount of scraps/waste produced, therefore the quantity of material available and the relative geographical position is known. If a hypothetical user of waste as a second raw material needs a specific recovery material, he will search for it in a given

and the economic estimation[19] of their possible recovery. This operation can be activated through the use of GIS.[20] In the research project conducted, the GIS has had several functions that have been studied and formalized in the procedures to be implemented in order to be able to use them fully and to exploit the related hypothetical further developments. The possible investigated uses are:

(a) Geographical contextualization of companies. Through the proposed system, companies—when they access the platform—record their position which is registered in a GIS. The GIS, acquiring the position of the companies, becomes useful for researches filtered by specific productive sector. A research of this type allows us to structure industrial synergies between companies (both in the same commodity sector, and in different sectors with common interests) spread over the territory, forming a sort of "virtual" industrial district that is not recognizable only by the territorial vocation.
(b) Geographical contextualization of scraps/waste. Through the proposed system, companies—when they access the platform—record the types of scraps/waste produced. They also automatically standardize the position of scraps/waste of their production chains This simple operation will allow users of the platform to identify scraps/waste on the territory, either by directly searching using the EWC code or by searching through the other reading keys. In this

territorial context (using the functions related to the search by territory). Therefore, this operation will allow to see in the territory who produces that waste, where and in what quantity.

[19]Once the distances of the materials to be enhanced have been established, the hypothetical user can estimate the cost of transporting the material, evaluating the maximum mileage (or distance in kilometers) that can be sustained in economic terms.

[20]A Geographic Information System, more commonly known as GIS, is a system created to be able to receive, store, process, analyze, manage and represent topographic data. In a simplified way, with a GIS system, it is possible to carry out statistical analysis and data management through databases attributed to maps. GIS systems are implemented with territorial information systems, which attribute information to objects present on the maps. GIS is therefore the software environment that allows the processing and manipulation of geo-referenced geometric data, which is stored in data structures of the DBMS (Database Management System) type. GIS is therefore a tool used to analyze spatial properties and potential relationships between objects and events. It is therefore a set of functions to map and analyze the phenomena that occur in space and time. Given that the analysis of spatial information is not a new concept, the contribution of the GIS was to overcome the barriers of paper-based thematic maps towards the digital field. This step allowed the manipulation of the information and their statistical evaluation in order to enable the resolution of complex problems.

regard the study[21] [11] conducted by the Polytechnic of Turin on C&D waste is a useful reference.

(c) Geographical contextualization of landfills or of treatment and recovery centers. As for companies, landfills and waste treatment centers can be mapped on GIS. They will be considered as places for temporary storage waiting for recovery and valorization. This function—if expanded and perfected—could also allow disposal centers to be profiled as "users", that would become a sort of real marketplace, in which different types of products could be found.

References

1. EUROSTAT (2016) Generation of waste by waste category, hazardousness and NACE Rev. 2 activity
2. ISPRA (2018), Rapporto rifiuti speciali. Edizione 2018, Ispra area comunicazione, Roma, 2018
3. Meyer B et al. (2011) Macroeconomic modelling of sustainable development and the links between the economy and the environment, GWS
4. Porter ME (1985) The competitive advantage: creating and sustaining superior performance. Free Press, NY, p 1985
5. Europe INNOVA (2012) Guide to resource efficiency in manufacturing: experiences from improving resource efficiency in manufacturing companies
6. Ellen MacArthur Foundation (2012) Towards the circular economy: economic and business rationale for an accelerated transition
7. Chertow M (2000) Industrial symbiosis: literature and taxonomy. Ann Rev Energ Environ 25:313–337
8. Trigilia C (2005) Sviluppo locale. Un progetto per l'Italia, Editori Laterza, Bari
9. Flora A (2008) Lo sviluppo economico: i fattori immateriali, nuove frontiere della ricerca. Franco Angeli Editore, Milano, pp 180–181
10. Pernice M, Mininni G (2008), Il sistema normativo e tecnico di gestione dei rifiuti, Ipsoa editore
11. Blengini GA, Garbarino E (2010) Resources and waste management in Turin (Italy): the role of recycled aggregates in the sustainable supply mix. J Clean Prod, Febbraio

[21] The study conducted by Politecnico di Torino analyzed the environmental implications deriving from the recovery of C&D waste within the administrative boundary of the Province of Turin. In particular, a study was carried out combining the GIS system and the LCA assessment method, using specific data from the territory and paying attention to the use of the territory, the transport system and the avoided landfills—crucial aspects for the sustainable planning and management of a territory. A GIS was therefore used to manage data and information on approximately 89 recycling plants, including data on the technological and physical/mechanical characteristics of the outgoing recycled aggregates. The LCA methodology was instead used to identify and quantify energy and environmental loads with reference to different possible scenarios outlined on the basis of factors such as: delivery distance, quality of recycled aggregates, local availability of natural aggregates and geographical coverage of the market demand. The purpose of the work was to test the eco-efficiency of the C&D waste recycling chain, especially with regard to the avoided impacts. The first results reported that over 90% of the indicators considered for the evaluations were positive (the avoided impacts were higher than the impacts induced by the activation of the recycling chain) and it was also estimated that the transport distance of the recycled aggregates should increase 2–3 times before the induced impacts overcome the avoided impacts. Source: Blengini and Garbarino [11] *Resources and waste management in Turin (Italy): the role of recycled aggregates in the sustainable supply mix*, Journal of Cleaner Production.

Chapter 6
Circular Economy and Sustainable Procurement: The Role of the Attestation of Conformity

Abstract The impact on the environment of the procurement processes has been considered, in the last years, as a critical issue to be managed in order to improve the sustainability performance of the organizations. At national and international levels, policies and regulations have been defined in order to push private and public organizations toward a more sustainable procurement process: from the European policies on green public procurement (GPP) to the international ISO standards on sustainable procurement there is an increasing attention to this subject. Within this framework of tools for the improvement of the sustainability in procurement, a key role is played by the circular economy approach and the use of recycled or reused materials and components is strongly encouraged. The demand for recycled or reused materials and components requires, on the other hand, criteria and methods to assess the real content of recycled matter in a building material or element; due to this reason new standards and regulations require that the recycled content of a product shall be declared under a scheme of attestation of conformity issued by a third party. The chapter present a review of the actual systems for the attestation of conformity of materials and products including a determined percentage of recycled materials.

Keywords Circular economy · Sustainable procurement · Environmental labels · Attestation of conformity

6.1 Environmental Protection, Green Procurement and Green Marketing

Public awareness about ecological and environmental issues had a significant increase from the beginning of the 1980s when topics as the ozone layer depletion, global warming and greenhouse effect or pollution—directly related to industrial manufacturing and anthropic activities—started to be associated with the purchase habits of public administration, private companies and consumers [1, 2]. This awareness has

This chapter is authored by Giancarlo Paganin, Politecnico di Milano, Department of Architecture and Urban Studies, Milan, Italy. e-mail: giancarlo.paganin@polimi.it.

M. Migliore et al., *Strategies for Circular Economy and Cross-sectoral Exchanges for Sustainable Building Products*, Springer Tracts in Civil Engineering,
https://doi.org/10.1007/978-3-030-30318-1_6

been formalized during the United Nations Conference on Environment and Development (UNCED) in 1992, when the international community recognised that current global patterns of consumption were not sustainable; excessive demands were being made on the planet's finite stock of resources and on the ability to manage the waste products of human activities. Chapter 4 titled "Changing Consumption Patterns" of the document "Agenda 21" issued during the Rio Conference was dealing with sustainable consumption principles and identified the strong importance of developing policies in this area in order to obtain "significant changes in the consumption patterns of industries, Governments, households and individuals" [3]. Governments were invited to work in order to promote and develop the consciousness of the sustainable consumption; in order to reach this consciousness, the environmental labelling has been identified as a success factor for the shift towards a more informed purchasing process (Table 6.1).

As defined by the Oslo Symposium in 1994, sustainable consumption and production (SCP) is about "the use of services and related products, which respond to basic needs and bring a better quality of life while minimizing the use of natural resources and toxic materials as well as the emissions of waste and pollutants over the life cycle of the service or product so as not to jeopardize the needs of further generations" (www.sustainabledevelopment.un.org).

Even if the Agenda 21 goals on sustainable consumption and production have remained unambitious for long time and Chap. 4 remained the least implemented of the various chapters of Agenda 21 [4], in the last few decades several activities have been developed in order to improve the sustainable consumption and production in three main areas: public procurement processes, business processes and consumer market. Public concern for environmental issues increased its importance in consumer's behaviour and in their potential role in the contribution to the protection of the environment [5] and according to this behavioural change, industry started

Table 6.1 Extracted from Agenda 21 on sustainable development

Role of labelling in sustainable consumption as per Agenda 21 Chap. 4
4.17. In the years ahead, Governments, working with appropriate organizations, should strive to meet the following broad objectives a. To promote efficiency in production processes … b. To develop a domestic policy framework that will encourage a shift to more sustainable patterns of production and consumption; c. To reinforce both values that encourage sustainable production and consumption …
4.21. Governments, in cooperation with industry and other relevant groups, should encourage expansion of environmental labelling and other environmentally related product information programmes designed to assist consumers to make informed choices
4.22. They should also encourage the emergence of an informed consumer public and assist individuals and households to make environmentally informed choices by a. Providing information on the consequences of consumption choices and behaviour so as to encourage demand for environmentally sound products and use of products b. Making consumers aware of the health and environmental impact of products, through such means as consumer legislation and environmental labelling

implementing new approaches to consumer market as, for instance, the green marketing. Green marketing was firstly developed in the 70s as a societal dimension of marketing to address limited availability of environmental resources, environmental impacts of conventional marketing and greening of the different aspects of traditional marketing but in the last 25 years reached a significant importance in the relation between customer and suppliers [6]. During the years, green marketing identified three types of tools, such as eco-label, eco-brand and environmental advertisement, with the objective to increase public awareness of green products attributes and characteristics [7]. The first two types of tool have been developed in parallel for the procurement processes of public administrations and business companies that intended to pursue the sustainable consumption and production objectives.

While the consumer market has developed methods and tools to improve consumer awareness on sustainable consumption, on the other hand the public administrations started considering the sustainable consumption as a critical issue in the public policies and green/sustainable procurement policies have been issued in several countries in the last decades. Nowadays [8, 9] almost all the OECD countries have been found having sustainable or green public procurement policies, tools and regulations in place. The reason why Governments and public decision makers are invited to develop sustainable procurement policies is that public spending may reach 15–30% of national GDP and it may represent an opportunity to drive also business purchasing towards innovation and sustainability [10].

UNEP defined 'sustainable procurement' as: "A process whereby organisations meet their needs for goods, services, works and utilities in a way that achieves value for money on a whole life basis in terms of generating benefits not only to the organisation, but also to society and the economy, whilst minimising damage to the environment" [10]. More recently another definition has been officially released by ISO International Organisation for Standardization that, in the standard ISO 20400: 2017 "Sustainable procurement—Guidance" defined sustainable procurement as: procurement (activity of acquiring goods or services from suppliers) that has the most positive environmental, social and economic impacts possible over the entire life cycle. Sustainable procurement involves the sustainability aspects related to the goods or services and to the suppliers along the supply chains [11].

The publication of the international standard ISO 20400:2017 shows that the interest of the industry and the researchers for the topic of sustainable procurement is strong and diffuse worldwide. The same interest for the topic of sustainable procurement is also involving scholars as it is shown by the continuous growth of the number of paper published on scientific journals: literature review [12] shows an exponential increase of number of papers in the last years when on a selected list of 15 journals the papers dealing with sustainable procurement have grown from 1 per year to 60 per year.

The definition of sustainable procurement is based on three dimensions: environmental, social/societal, economic. The three dimensions may be summarized, from the industry and business point of view as [13]:

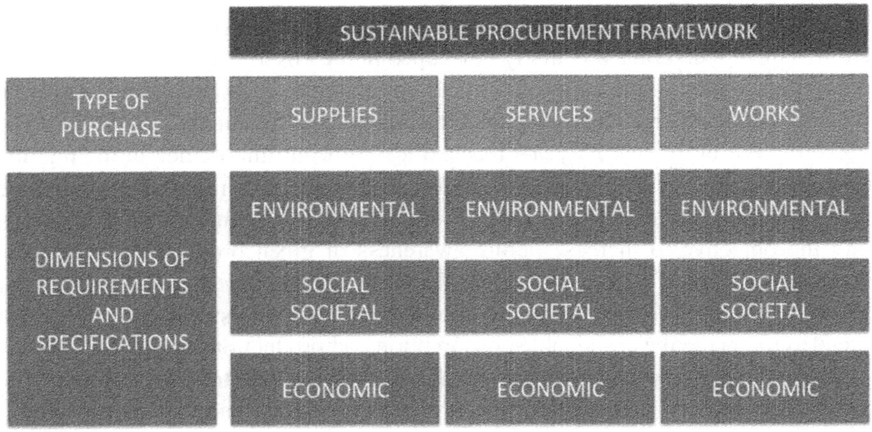

Fig. 6.1 Framework of sustainable procurement process

- economic sustainability deals with striving for profit and growth;
- environmental sustainability means to act in terms of producing and consuming with minimized harm to the environment;
- social sustainability concerns producing and consuming with maximized benefits for humankind and society in general.

The three dimensions of sustainability are declined, within the sustainable procurement process, with respect to three types of purchase: goods/supplies, services, works as shown in Fig. 6.1.

From a practical point of view, in order to develop tools and methods to require and monitor the sustainable performances of procurement process, the sustainable procurement framework can be broken down into sub levels or tiers dealing with more specific aspects of the three dimensions. A possible example of breakdown is presented in Fig. 6.2.

Focusing the attention on the environmental dimension of the sustainable procurement process, it is possible to identify a first topic that has been developed from mid 90 s: the green public procurement.

Green Public Procurement (GPP) is defined in the EU as "a process whereby public authorities seek to procure goods, services and works with a reduced environmental impact throughout their life cycle when compared to goods, services and works with the same primary function that would otherwise be procured" [14]. European GPP [15–17] policy falls within a range of various action developed by several Governments to encourage sustainable procurement as, for instance:

- the Environmentally Preferable Purchasing of the Federal Government of the United States. "Environmentally preferable" is defined in Executive Order 13101 of 1998 "Greening the Government through Waste Prevention, Recycling and Federal Acquisition" to mean products or services that "have a lesser or reduced effect on human health and the environment when compared with competing products or

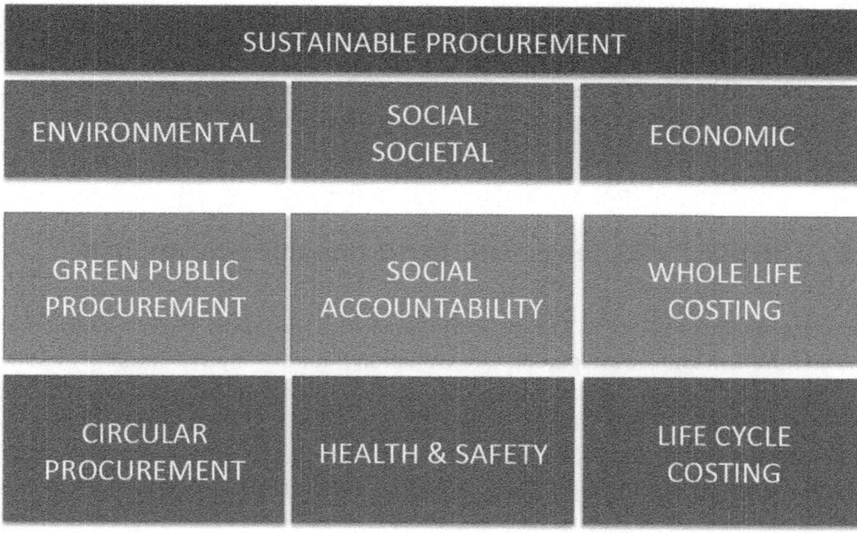

Fig. 6.2 Example of breakdown of sustainable procurement issues

services that serve the same purpose. This comparison may consider raw materials acquisition, production, manufacturing, packaging, distribution, reuse, operation, maintenance or disposal of the product or service";

- the Green Public Procurement program started in 2001 in Japan with the "Act on Promotion of Procurement of Eco-Friendly Goods and Services by the State and Other Entities (Act on Promoting Green Purchasing). The program is mandatory for Government agencies and public institutions, is encouraged for Local governments and suggested as voluntary decision to private companies and citizens; the purpose stated in article 1 of the Act is "to establish a society that can develop sustainably with reduced environmental loads, by providing for necessary matters and for promoting a shift of demand to eco-friendly goods, etc., including the promotion of the procurement of eco-friendly goods, etc. by the State, incorporated administrative agencies, etc.";
- the Korean Law "Act on the Promotion of the Purchase of Environment-friendly Products", enforced in July 2005, which aimed to encourage the purchasing of environment-friendly products and services by means of mandating public agencies to buy environment-friendly products and services and supporting industry and household-level green consumption [18];
- since 2007, the central and provincial governments in China are required to prioritize environment-friendly products listed in a green product inventory. Products purchased by Public Administration must meet the environmental protection and energy saving standards set by the State Environmental Authority. In the beginning of the third millennium, the Chinese Government has recognized that green purchasing could contribute significantly to national environmental and sustainable development [19]. Within the general framework of public procurement law of

2002, two sets of goods and services have been identified to develop a green public procurement policy: a "Government Procurement List on Energy Saving Products" in 2004 and a "Government Procurement List for Environmentally Labelled Products" in 2007 have been prepared and applied on public purchasing.

6.2 Circularity Requirements in Sustainable Procurement

Since the publication in 1987 of the well known "Report of the World Commission on Environment and Development: Our Common Future"—more commonly known as Brundtland Report from the name of Norwegian Prime Minister Gro Harlem Brundtland head of World Commission on Environment and Development—the idea of closing the loop by recycling materials and energy has been an important issue of sustainable procurement. The development of the different GPP policies has therefore considered since the beginning the fact that recycle and reuse of materials and energy are key success factors for the achievement of sustainability targets.

As any procurement process, GPP shall clearly define and apply criteria to require the supply and to check if the supply is in compliance with the expressed requirements. The activity of criteria specification is normally performed during the preparation stage and the specification stage of the procurement process and the criteria are later used in the sourcing stage and utilisation stage to select supplier and to assess quality of supply [15]. In order to be effective and replicable over a fragmented set of purchasing entity—represented by public administrations at different scales—the different GPP systems are supported by criteria that can be used by single purchasing entity in order to comply with the GPP general policy. Table 6.2 shows an example of the categories for which GPP criteria have been developed in different GPP systems.

One of the key aspects in the set up of environmental criteria is related to the general objective of achieving a better use of natural resources and minimize wastes as emphasized by Agenda 21 asking Governments to "make a concerted effort to reduce the generation of wastes and waste products by encouraging recycling in industrial processes and at the consumed level". The major part of various environmental criteria developed over time includes specifications dealing with a minimum percentage of recycled matters in the goods to be supplied within the framework of GPP. For instance, the UE environmental criteria "EU GPP Criteria for Office Building Design, Construction and Management" foresees that "The procurer shall award points to tenderers that achieve greater than or equal to 15% by value of recycled content and/or by-products for the sum of the main building elements". Another example of the same criteria is shown in Table 6.3.

The aspect of GPP related with the sustainable use of natural resources with the support of recycling processes is perfectly in line with the vision of circular economy and it starts being considered in a broader perspective of supply chain management. The integration of circular economy principles within the supply chain management would begin to extend the boundary of supply chain management by reducing the

Table 6.2 Example of products/services categories for the application of environmental criteria within GPP or SPP framework. (1) Source www.ec.europa.eu/environment/gpp/eu_gpp_criteria_en.htm (2) Japan Ministry of Environment (3) US General Services Administration (GSA) https://sftool.gov/greenprocurement

EU GPP (1)	Japan GPP (2)	USA (3)
Cleaning products and services	Printing and copying papers	Appliances
Computer and monitors	Copiers, printers, facsimile machines	Biomedical equipment and supplies
Copying and graphic paper	Toilet paper	Building finishes
Electrical and electronic equipment in the health care sector	Tissue paper	Building furnishings
Electricity	Personal computers	Cafeteria products
Food and catering services	Refrigerators	Cleaning products
Furniture	Stationery and office supplies	Construction materials
Gardening products and services	Washing machines	Contracted printing products
Imaging equipment	Lighting equipment	Doors and windows
Office buildings design, construction and management	Automobiles	HVAC/Mechanical
Paints, varnishes and road markings	Air conditioners	Landscaping products
Road design, construction and maintenance	Office furniture	Lighting and ceiling fans
Road lighting and traffic signals	Television sets	Lube, oil, hydraulic fluid, and grease
Road transport	Uniforms and workwear	Miscellaneous
Sanitary tapware	Printing services	Non-paper office products
Textiles	Hotels and inns	Office electronics
Toilets and Urinals	Toilet facilities	Paper office products
Waste water infrastructure	Food (processed food)	Park and recreation
Water-based heaters	Tradable green certificate	Personal care
	Truckload shipping	Plumbing systems
	Electricity	Refrigeration systems
		Renewable energy
		Roadway construction
		Shipping, packaging and packing supplies
		Special construction products
		Traffic control

(continued)

Table 6.2 (continued)

EU GPP (1)	Japan GPP (2)	USA (3)
		Vehicles and vehicle products
		Cafeteria and food services
		Electronic equipment leasing
		Fleet maintenance

Table 6.3 Example of specification for minimum recycled content of building insulation as per the US EPA's recovered materials advisory notice within the framework of comprehensive procurement guideline (CPG) program

Product	Material	Postconsumer content (%)	Total recovered materials content (%)
Rock wool	Slag	–	75
Fiberglass	Glass cullet	–	20–25
Cellulose loose-fill and spray-on	Postconsumer paper	75	75
Perlite composite board	Postconsumer paper	23	23

need of virgin materials which could increase the circulation of resources within supply chains systems [20].

In the last years the number of papers concerning the theme of circular supply chain has increased significantly and recently definitions for the topic has been proposed as, for instance the one that defines circular supply chain as: "the integration of circular thinking into the management of the supply chain and its surrounding industrial and natural ecosystems. It systematically restores technical materials and regenerates biological materials toward a zero-waste vision through system-wide innovation in business models and supply chain functions from product/service design to end-of-life and waste management, involving all stakeholders in a product/service lifecycle including parts/product manufacturers, service providers, consumers, and users" [16].

From the point of view of GPP policies, the inclusion of circular economy principles in green public procurement processes led to define the idea of Circular Public Procurement (CPP) as "an approach to greening procurement which recognises the role that public authorities can play in supporting the transition towards a circular economy. Circular procurement can be defined as the process by which public authorities purchase works, goods or services that seek to contribute to closed energy and material loops within supply chains, whilst minimising, and in the best case avoiding, negative environmental impacts and waste creation across their whole life-cycle" [21].

The specification of circular requirements, namely the minimum percentage of recycled content in materials or components, in procurement processes is, as any other technical specification, related to methods and tools that are needed from the clients and specifiers in order to assess quality of supply, i.e. the compliance between the supply and the requirements. Procurement practitioners are not necessarily experts in environmental issues or requirements and, therefore, it may be not effective to rely on the direct verification of the supplied products by procurement operators. One of the possible approaches to this kind of assessment is to entrust to external subjects or third parties the task to verify and declare the fulfilment of the concerned requirements. This approach led over the years to the development and the dissemination of environmental labels of different nature.

6.3 Attestation of Conformity and Product Labeling in Circular Economy Processes

Labelling of products concerning their—whole or partial—environmental performance is a tool that since around 40 years is used by industry to inform consumers who started to develop an environmental awareness. The first "ecolabel" introduced on the market has been the German "Blue Angel" that was created in 1978 by the German Federal Ministry for Interiors [22]. Following the Blue Angel label, several experiences of ecolabels and environmental labels have been developed all around the world and nowadays more than 450[1] ecolabel programs are implemented on several types of products or services. In 2010 a survey [23] highlighted the main characteristics of a sample of more than 300 ecolabels: almost 60% of them were managed by non-profit organization and 8% directly by Governments; 64% were associated to a third party assessment and 27% to a second party assessment; labelling programs run by non-profits tended to have more conformity assessment requirements such as site visits, audits and third party certifications than privately-led schemes; ecolabels are often worldwide known and more than 60% were not limited to one country in their scope; concerning awarding of the label more than 70% of the programs were pass or fail type—the ecolabel is either awarded or not awarded if a certain threshold of performance is met—and less than 20% of the programs were tiered with a specific rating system [24, 25].

Environmental labelling can be defined as "making relevant environmental information available to the appropriate consumers. Environmental labelling is the practice of labelling products based on a wide range of environmental considerations (e.g., hazard warnings, certified marketing claims, and information disclosure labels)" [26]. The environmental labels represent a communication tool to inform purchaser and customer about the conformity with specific requirements as, for instance, the above mentioned GPP environmental criteria [27]. Conformity assessment is defined as the "demonstration that specified requirements relating to a product, process, system,

[1] Source: www.ecolabelindex.com.

person or body are fulfilled" [28] and it represents the basis for issuing environmental labels. Depending on the independence of the person or body carrying out the conformity assessment, there may be three configurations:

- first-party conformity assessment: is the assessment performed directly by the subject (person or organization) that provides the product or service under exam. The subject has a specific interest in the product and therefore is not independent in its assessment;
- second-party conformity assessment: is the assessment performed by a subject (person or organization) that has a user interest in the product or service under exam. Again, the subject has an interest (for instance because he's the buyer of the product or it's an association of manufacturers producing the type of product under exam) in the product and therefore is not independent in its assessment;
- third-party conformity assessment: is the assessment performed by a subject (person or conformity assessment body) that is independent of the organization or person providing the product, and of user interest in that product. The third party is supposed to be independent in its assessment due to the absence of any interest in the object under examination.

Relationship between conformity assessment and labels should always be kept in consideration by their users: a label normally expresses that a product or system comply with specific requirements and the consumer or buyer should know which requirements have been fulfilled and the way in which the conformity with requirements has been assessed. This knowledge is normally quite difficult to acquire by a consumer and therefore it is important that the market could consider the labels reliable and this reliability is practically entrusted to the issuer independence.

Environmental labelling programs can, in general, be characterized by three main key topics:

- Independence: the independence of the subject issuing the label is an important aspect related to the reliability of the label (see above). Consumer sometimes are not totally aware about the difference between a first party or third party label and they cannot understand the presence of a potential conflict of interest in the first party labelling;
- Compulsoriness: environmental labels may be required on a mandatory basis (by Law) or may be applied on a voluntary basis by the manufacturer in order to disclose the information about environmental performance of its product or service. Mandatory labelling programs typically include hazard or warning labels.[2] Voluntary labels are typically sending a positive or at least a neutral message as it is unlikely that manufacturers voluntarily disclose negative information about their products;

---

[2]The probably best-known compulsory labelling system is the international "Globally harmonized system of classification and labelling of chemicals (GHS)" proposed by the United Nations. At European regional level the GHS is the basis for the Classification, Labelling and Packaging (CLP) Regulation ((EC) No 1272/2008).

- Type of message: the message transmitted with the label emphasizes the environmental performances of the product or service that may be critical to guide the decisions in the purchasing process by consumers or by organizations. Positive labelling typically is used to communicate that labelled products possess one or more environmentally preferable attributes. Negative labelling warns consumers about the harmful or hazardous ingredients contained in the labelled products. Neutral labelling programs simply summarize environmental information about products that can be interpreted by consumers as part of their purchasing decisions [24] (Fig. 6.3).

Within the large number of environmental labelling programs internationally applied, a key role is played by the system proposed by the international standard ISO 14020:2000 "Environmental labels and declarations—General principles". The standard has been prepared starting from the consideration that "the overall goal of environmental labels and declarations is, through communication of verifiable and accurate information, that is not misleading, on environmental aspects of products and services, to encourage the demand for and supply of those products and services that cause less stress on the environment, thereby stimulating the potential for market-driven continuous environmental improvement". The basic requirements defined in the ISO 14020 standard and to be considered in preparing an environmental labelling program are:

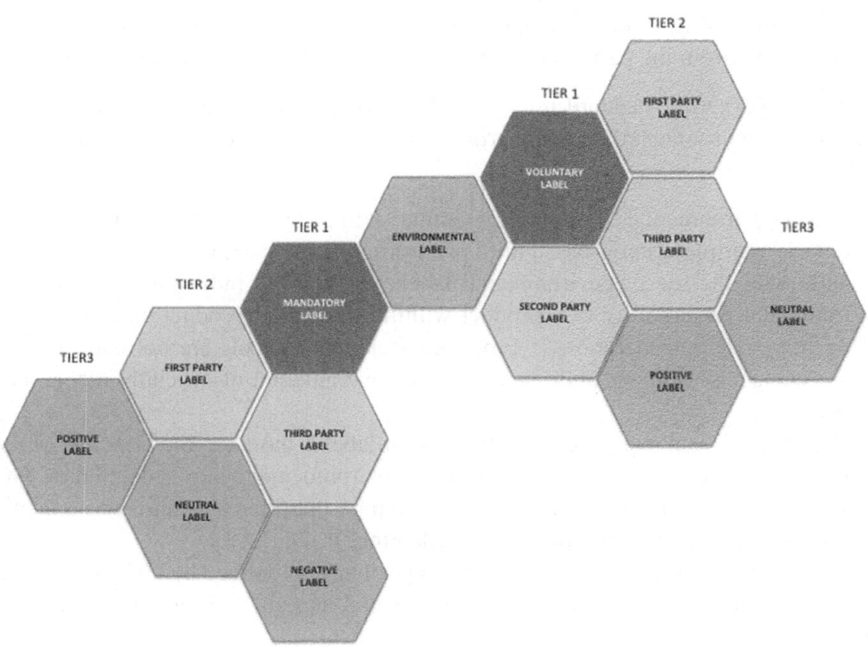

Fig. 6.3 Breakdown of different environmental labelling systems

- environmental labels and declarations shall be accurate, verifiable, relevant and not misleading;
- procedures and requirements for environmental labels and declarations shall not be prepared, adopted, or applied with a view to, or with the effect of, creating unnecessary obstacles to international trade;
- environmental labels and declarations shall be based on scientific methodology that is sufficiently thorough and comprehensive to support the claim and that produces results that are accurate and reproducible;
- information concerning the procedure, methodology, and any criteria used to support environmental labels and declarations shall be available and provided upon request to all interested parties;
- the development of environmental labels and declarations shall take into consideration all relevant aspects of the product life cycle;
- environmental labels and declarations shall not inhibit innovation which maintains or has the potential to improve environmental performance;
- any administrative requirements or information demands related to environmental labels and declarations shall be limited to those necessary to establish conformance with applicable criteria and standards of the labels and declarations;
- the process of developing environmental labels and declarations should include an open, participatory consultation with interested parties. Reasonable efforts should be made to achieve a consensus throughout the process;
- information on the environmental aspects of products and services relevant to an environmental label or declaration shall be available to purchasers and potential purchasers from the party making the environmental label or declaration;

ISO 14020:2000 standard refers to other standards to define different ways of managing environmental labelling programs introducing three types of environmental labels:

- Type I environmental labelling programme: defined by ISO 14024:2018 as a "voluntary, multiple-criteria-based third party programme that awards a licence which authorizes the use of environmental labels on products indicating overall environmental preferability of a product within a particular product category based on life cycle considerations". Type I environmental labels are normally known as "ecolabels" and are often managed by governmental of officially recognized bodies[3];
- Type II environmental labelling: this type of label is based on the principle of the self declared environmental claim (where environmental claim is defined by ISO 14021:2016 as "a statement, symbol or graphic that indicates an environmental aspect of a product, a component or packaging");
- Type III environmental declaration: this type of label is defined by ISO 14025:2006 as a "claim which indicates the environmental aspects of a product, providing

[3]There are several examples of Type I labelling programs as: European Ecolabel, Eco Mark Japan, Nordic Swan, Blue Angel, China Environmental Labelling, Eco Mark India, Korean Ecolabel, ….

Table 6.4 Main characteristics of ISO 14020 environmental labelling types

Type of label	Reference environmental criteria	Minimum acceptance threshold	Issuer	User
Type I	Multiple	Yes	Third party	Consumer
Type II	Single/multiple	No	First party	Consumer
Type III	Multiple	No	Third party	Business

quantified environmental data using predetermined parameters and, where relevant, additional environmental information" (Table 6.4).

Type I and Type III labels are based on their ability to measure the environmental impacts of the labelled products or services over their life cycle and; in order to reach that goal, in a Life Cycle Assessment (LCA) perspective, they need to have available a set of measuring tools i.e. an appropriate metric. In order to transfer and apply the labelling programmes to the circular economy scope there is the need to define a metric and a set of circularity indicators on the basis of which it is possible to assess the circularity performances of products and services and labelling them according to those assessment results.

In the last years several studies and project have been started on order to define circularity metrics and indicators [29]. Circularity metric may be used to assess circularity performances of companies and, in the same time, they can be the basis for product labelling programs to inform consumer choices and to support procurement activities between companies or within the public sector. Circularity metrics can potentially function to push for a transition to a circular economy in which it can allow customers to elicit demands for products with a higher degree of circularity and encourage manufacturers to engage in material recirculation activities [30].

6.4 Conclusions

Procurement and purchasing processes may play a key role in the development of circular economy approach [31, 32], triggering interest of manufacturing companies and service providers in reshaping their business models to compete with the new market requests. In this direction, consumers and procurement practitioners need to be supported in their choices by reliable information about the environmental and circularity performances of products and service.

Environmental labels have been developed on a large scale in the last decades in order to support the transition towards a sustainable production and consumption system as envisaged by Agenda 21 on Sustainable Development. They have been included in almost all the Sustainable Public Procurement or Green Public Procurement processes in order to supply procurement practitioners with a simple tool to fulfil their duty to procure items that are environmentally preferable. Very often environmental labels include in their requirements the assessment of the content of recycled matters and the recyclability or reusability of products, in line with the "closing the loop" principle characterizing circular economy processes.

Labelling programs for circular economy processes have been developed recently[4] in order to provide consumers and buyers with tools useful to consider the circularity requirements in their purchases. To increase the variety and spread of circularity labels it is however necessary to supply labelling program operators with appropriate metrics and circularity indicators in order to develop the conformity assessment schemes necessary for the issue of labels.

References

1. Schlegelmilch B, Bohlen G, Diamantopoulos A (1996) The link between green purchasing decisions and measures of environmental consciousness. Eur J Mark 30(5):35–55
2. Yan YK, Yazdanifard R (2014) The concept of green marketing and green product development on consumer buying approach. Global J Commer Manag Perspect 3(2):33–38
3. UNEP (2001) Consumption opportunities. Strategies for change. A report for decision-makers, UNEP, Geneva
4. Fuchs DA, Lorek S (2004) Sustainable consumption. Political debate and actual impact, SERI Background Papers 4
5. Lerch M (2017) Challenges of globalization: new trends in international politics and society. Taylor & Francis
6. Kumar P (2016) State of green marketing research over 25 years (1990–2014). Market Intell Plann 34(1):137–158
7. Delafrooz N, Taleghani M, Nouri B (2014) Effect of green marketing on consumer purchase behavior. QScience Connect 5 http://dx.doi.org/ https://doi.org/10.5339/connect.2014.5
8. OECD (2017) Government at a glance 2017. OECD Publishing, Paris
9. Walker H, Mayo J, Brammer S, Touboulic A, Lynch J (2012) Sustainable procurement: an international policy analysis of 30 OECD countries. In: 5th International Public Procurement Conference (pp. 3556–3582)
10. UNEP (2012) Sustainable public procurement implementation guidelines. UNEP, Paris, p 2012
11. Walker H, Miemczyk J, Johnsen T, Spencer R (2012) Sustainable procurement: past, present and future. J Purchas Supply Manag 18(4)
12. Touboulic A, Walker H (2015) Theories in sustainable supply chain management: a structured literature review. Int. J. Phys. Distr. Logist. Manag. 45(1/2):16–42
13. Fischer D, Brettel M, Mauer R (2018) The three dimensions of sustainability: a delicate balancing act for entrepreneurs made more complex by stakeholder expectations. J Bus Ethic 1–20
14. European Commission (2008) COM (2008) 400, published on 16 July 2008, Communication "Public procurement for a better environment"
15. Erdmenger C (2017) Buying into the environment: experiences, opportunities and potential for eco-procurement. Taylor & Francis
16. European Commission (2016) Buying green! A handbook on green public procurement, 3rd edn. https://doi.org/10.2779/246106
17. OECD (2013) Mapping out good practices for promoting green public procurement. OECD meeting of Leading Practitioners on Public Procurement, Paris, p 2013
18. OECD (2008) Eco-innovation policies in the Republic of Korea. OECD, Environment Directorate

[4]As example the experiences of "Cradle to Cradle Certified™" and "Remade in Italy" labels can be considered.

19. Ho L, Dickinson NM, Chan GY (2010) Green procurement in the Asian public sector and the Hong Kong private sector. Nat Resour Forum 34(2010):24–38
20. Farooque M, Zhang A, Thürer M, Qu T, Huisingh D (2019) Circular supply chain management: a definition and structured literature review. J Clean Prod 228(2019):882–900
21. European Commission (2017) Public procurement for a circular economy. Good practice and guidance
22. United Nations Office for Project Services (UNOPS) (2009) A guide to environmental labels for procurement practitioners of the united nations system. UNOPS publications
23. World Resources Institute (2010) Global ecolabel monitor. Towards transparency
24. Gruère G (2013) A characterisation of environmental labelling and information schemes. OECD Environment Working Papers, No. 62, OECD Publishing, Paris
25. Horne RE (2009) Limits to labels: the role of eco-labels in the assessment of product sustainability and routes to sustainable consumption. Int J Consum Stud 33(2009):175–182
26. US Environmental Protection Agency (US EPA) (1998) Environmental labeling issues, policies, and practices worldwide, EPA 742-R-98-009, December 1998
27. Prag A, Lyon T, Russillo A (2016) Multiplication of environmental labelling and information schemes (ELIS): implications for environment and trade. OECD Environment Working Papers, No. 106, OECD Publishing, Paris. https://doi.org/10.1787/5jm0p33z27wf-en
28. ISO 17000:2004 "Conformity assessment. Vocabulary and general principles"
29. The Ellen MacArthur Foundation (2015) Circularity indicators. An approach to measuring circularity, May 2015
30. Linder M, Sarasini S, Loon P (2017) A metric for quantifying product-level circularity. J Ind Ecol 21:545–558. https://doi.org/10.1111/jiec.12552
31. Mansvelt J (2011) Green consumerism: an A-to-Z guide. Sage reference series on green society, SAGE Publications
32. UN (2012) Low carbon green growth roadmap for Asia and the Pacific, United Nations Publications

Chapter 7
Crossing the Boundaries: From Agriculture and Livestock to the Building Industry

Abstract The increase in the world population in the last decades has led to a greater demand for agricultural and livestock processes. Wastes from these productions are becoming more and more important and their impacts on the environment may be of major level. On the other hands the chance to re-use agricultural and livestock byproducts and wastes offers the opportunity to rely on a renewable resource that can be controlled and planned in its availability. Since several years there is a research ongoing in order to find methods and processes for the recycle or reuse of agricultural wastes but so far their use still represents a minor resource for the construction industry. The chapter presents, also with the support of representative case studies, the actual and future trends in the research for the creation of cross sectoral stream of matter from agriculture and livestock to the construction industry.

Keywords Circular economy · Agricultural scraps · Livestock scraps · Construction products

7.1 Wastes from Agriculture and Livestock: Environmental Issues

In the last 2000 years the world population has grown from around 0.3 to 7.5 billion people [1] and, as a consequence, the production of the agricultural sector has grown proportionately along with the corresponding generation of wastes and by-products. Furthermore, as non-secondary issue, the increase in agricultural production and livestock has resulted in a significant increase in the impacts generated on the environment by these activities, especially with regard to the emission in the atmosphere of climate-changing gases, namely greenhouse gases (GHG). In order to better represent the growth of the agricultural sector and its impact on global GHG emissions, two specific areas are below analysed: cattle breeding and rice cultivation. For these two sectors the available data (FAOSTAT) show an important growth in the last 50 years:

This chapter is authored by Giancarlo Paganin, Politecnico di Milano, Department of Architecture and Urban Studies, Milan, Italy. e-mail: giancarlo.paganin@polimi.it.

M. Migliore et al., *Strategies for Circular Economy and Cross-sectoral Exchanges for Sustainable Building Products*, Springer Tracts in Civil Engineering,
https://doi.org/10.1007/978-3-030-30318-1_7

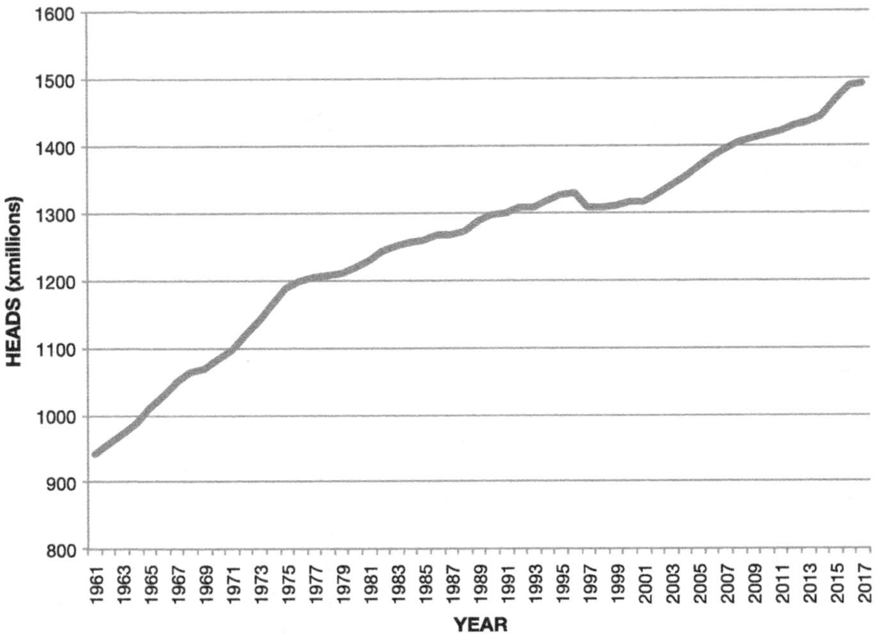

Fig. 7.1 World stock cattle trend (FAOSTAT)

- as regards livestock, the stock of animals increased globally from 1 billion to 1.5 billion animals (+50% in 50 years) (Fig. 7.1) even if in Europe the number of cattle has declined from around 200 million in the 60 s to the current 120 million head (source FAOSTAT); if sheep's, goats and poultry birds are also considered, the number of farm animals reaches over 25 billion [2];
- as regards rice production, in the same reference period of the last 50 years, production almost tripled, from 250 million tons to about 750 million tons per year (Fig. 7.2). Even in Europe, rice production has increased significantly from less than two million tons in 1960 to 4 million tons in 2017 (source FAOSTAT).

This important and continuous growth in agricultural production and the number of livestock, is leading to a significant increase in environmental loads related to the management of waste or by-products of these processes, particularly with regard to greenhouse gas emissions expressed in CO_2 equivalent. Figures 7.3 and 7.4 show the trend in greenhouse gas production in the whole agriculture and in cattle breeding and rice cultivation estimated by FAO in accordance with the indications of the IPCC guidelines.

This trend in the agricultural and livestock sector is part of a pattern of global climate-changing gas emissions (Fig. 7.5) which sees on the one hand a significant weight of the agricultural sector and, on the other, it sees the other sectors responsible for emissions—industry, energy and transport—engaged in specific emission

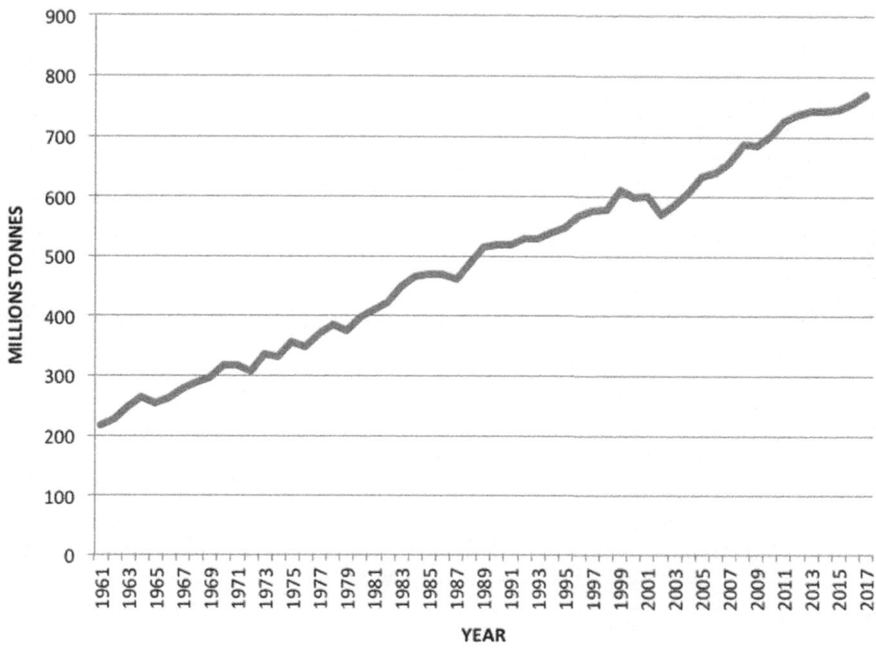

Fig. 7.2 World rice production trend (FAOSTAT)

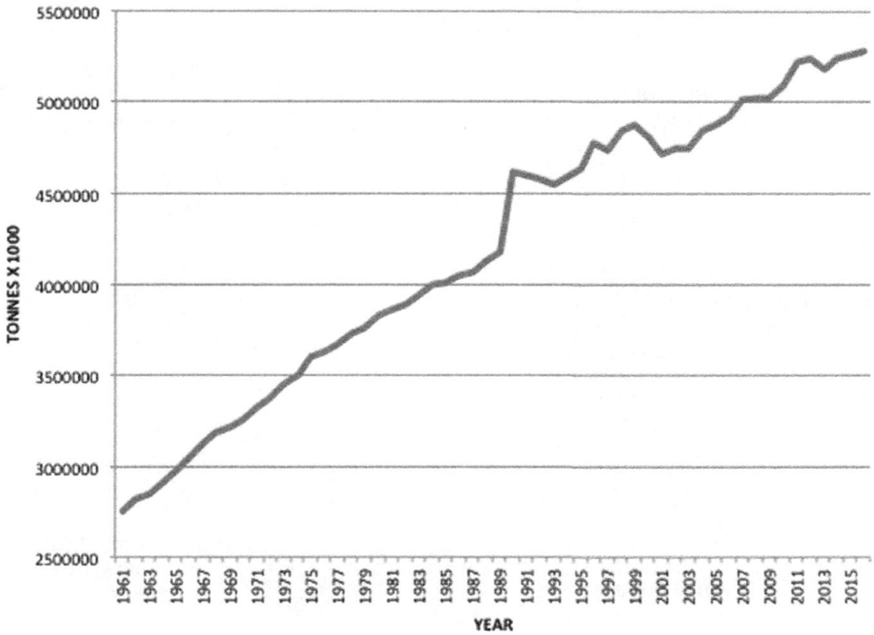

Fig. 7.3 Trend of yearly CO_2 eq emissions due agriculture. *Source* FAOSTAT

Fig. 7.4 Trend of yearly CO_2 emissions from manure management and rice cultivation. *Source* FAOSTAT

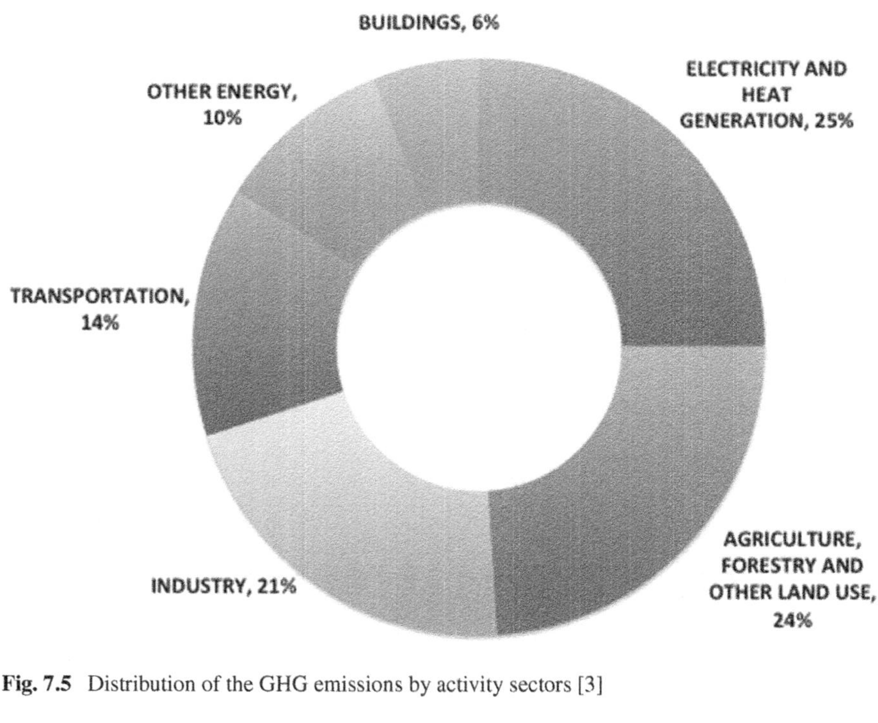

Fig. 7.5 Distribution of the GHG emissions by activity sectors [3]

reduction actions (for example, the ETS system introduced in Europe[1]). It is therefore important that the agricultural sector identifies lines of action for improving its performance in terms of greenhouse gas emissions; since emissions are linked for a significant part to the methods of managing agricultural production waste, it is reasonable to think that one of the global GHG emission reduction strategies could be to implement a mechanism for the recovery of renewable material resources in a perspective of circular economy.

In the global scenario of greenhouse gases emissions, agriculture is responsible for about half of the global anthropogenic methane emissions and methane impacts about twenty times more than carbon dioxide in its warming action. Livestock account for about a quarter of total methane emissions as consequence of two processes: enteric fermentation—related to the number of heads in the livestock—and manure management. The production of manure is projected to rise by about 60% by 2030 and greenhouse gases emissions from livestock are likely to increase by the same proportion. Rice cultivation is the other main agricultural source of methane, accounting for about a fifth of total anthropogenic emissions. The area used for irrigated rice is projected to increase by about 10% by 2030 (www.fao.org).

7.2 Present and Future Opportunities for Cross Sectoral Stream to the Building Industry

The activities related to agriculture and breeding have always been an area of experimentation for the development of ante litteram circular economy processes: the reuse of waste and by-products from cultivation and breeding has been developed since ancient times for the management of a large quantity of organic matter that was made available with substantially regular cycles linked to the harvest seasons or the daily care of animals [4, 5]. The use of crop residues and manure for the production of various artefacts—from the combustible tiles to bricks or plaster—has been a significant element for the development of civilization in many areas and has left us with important suggestions to evaluate the opportunities to consider the agriculture and livestock sectors as important suppliers of material for circular economy processes.

The recovery of resources deriving from waste and by-products from agriculture and breeding appears to be an important opportunity to pursue even in relation to the available quantity of resources: in the European Union alone, an estimated 95 million tonnes of waste are produced annually (Table 7.1)—a value that has steadily increased over the last 10 years—generated by agricultural and forestry activities (Eurostat source) of which about 15 million tons of manure. Worldwide, the amount of waste generated by agricultural and forestry activities is estimated a little less than one billion tonnes per year [6].

[1] Directive 2003/87/EC of the European Parliament and of the Council of 13 October 2003 establishing a scheme for greenhouse gas emission allowance trading within the Community and amending Council Directive 96/61/EC.

Table 7.1 Trend in agricultural waste generation in Europe (EU/28) and in major countries (tonnes/year)

	2010	2012	2014	2016
European Union-28	88.280.000	85.670.000	88.220.000	94.070.000
Belgium	4.822.033	4.979.571	5.654.085	6.726.241
Germany	12.932.869	14.086.695	15.011.037	15.623.841
Spain	9.763.484	8.297.358	8.058.640	9.135.749
France	9.405.994	11.281.262	11.304.163	11.812.213
Italy	9.489.613	5.785.758	6.929.296	7.852.188

Source EUROSTAT

Over the last few decades, numerous attempts and experiments have been developed for the reuse of agricultural production waste; in general, some consolidated treatment and reuse/recycling chains of agricultural waste can be identified [7] which mainly concern:

- Biogas/biofuel: Biogas is produced from the anaerobic fermentation of organic materials in presence of microorganisms and under controlled conditions. It is a mixture of gases—mainly methane—that results from anaerobic fermentation of organic matter by the action of methanogens bacteria. There are generally two types of methanogens bacteria used in digestion plants: Methanosarcina and Methanosaeta. Biofuel is produced from plant biomass and fatty acids and it can have different types of products such as ethanol, methanol, butane and bio-diesel.
- Briqueting: agricultural waste burns very quickly due to shape and density factors, so a widely used process is to increase their density (reducing the combustion speed) by pressing them in the form of pellets or briquettes. In this way, as well as improving the behaviour as an energy source, the ease of their transport is increased since, for the same volume, more mass and therefore more thermal energy can be transported.
- Animal fodder: this chain of recycling of agricultural production waste is still widespread but, due to the characteristics of the products used, it often requires special treatments to make the products usable also because of their low nutritional value. The most frequent are mechanical treatments (shredding, grinding, immersion in water,…) and chemicals (addition of urea and ammonia).
- Composting: Composting is a process based on the aerobic decomposition of organic materials by microorganisms under controlled conditions of temperature, moisture and oxygen availability. The process improves organic waste and kills pathogen organisms in the organic waste product in order to obtain a material (called compost) basically used as additive to soil to improve quantity of humus and nutrients.
- Bio based materials: bio-based materials or bio-based products can be defined as "product wholly or partly derived from biomass" where biomass is "material of biological origin excluding material embedded in geological formations and/or

fossilized" [8]. Bio-plastics, bio-lubricants, bio-insulation are more and more frequently used as common terms in procurement processes.

Due to the interest in the potential recycling and re-use of resources represented by waste from the agricultural sector, studies on the subject are becoming increasingly numerous and deepen various possible supply chains—with frequent attention to reuse in the construction sector—for the reuse of agricultural waste [7, 9–12] that can be used for example for the production of: particle boards, insulation boards, wall panels, printing paper and corrugating medium, roofing sheets, fuel, binder, fibrous building panels, bricks, acid proof cement, coir fibre, reinforced composite, polymer composites, cement board.

Most of the potential applications of agricultural waste look at construction as a sector of interest also due to its following peculiarities:

- it is a productive sector that absorbs a large quantity of material to make finished products (buildings and construction works) that are bulky, heavy and by their nature destined to a service life, in the order of several tens of years as minimum, not comparable with the life of other industrial products;
- it is a sector that, also due to the permanence of craftsmanship or semi-artisan work, allows the use of raw materials and semi-finished products with a quality standard that shall not always, especially for some types of tasks, ensure a high level as in high-tech sectors such as automotive or aerospace;
- is a sector in which the development policies for new projects, especially in countries most industrialized and sensitive to environmental issues, foresee environmental requirements which often includes purchase specifications and prescription of materials with a minimum content of recycled material.[2]

With respects to this great range of possibilities for the activation of waste streams from the agriculture sector to the construction sector, it is important that initiatives concerning the circular economy they consider not only the possibility of reusing a secondary raw material but also the benefits in environmental, as well as economic, terms that such reuse may involve. It is important to adequately investigate the information relating to the characteristics and quantities involved in the waste management process and to assess the environmental and economic aspects and impacts associated with managing the waste in question. Awareness of the environmental problem linked to waste management is the basic condition for a correct assessment of the feasibility and development opportunities of the circular economy project. The following paragraph intends to present two possible approaches to the reuse and recycling of agricultural production waste in the construction sector, highlighting the assessments regarding the environmental benefits obtainable by innovating the current waste and waste management processes.

[2]In this sense examples are given by the rewarding criteria introduced by voluntary environmental certification protocols such as LEED or BREEAM for material with recycled content in the projects or by the requests introduced in some countries within the framework of green public procurement policies (see also "Public procurement for a better environment" COM (2008) 400, published on 16 July 2008).

7.3 Case Studies: Rice and Manure Management Recycling Processes in Building Industry

7.3.1 *Rice*

The above mentioned growth of rice production involves a positive impact in terms of greater availability of food for the growing world population but should also be assessed in relation to the environmental aspects related to the management of by-products and wastes from rice processing, particularly as regards the husk, the bran and the rice straw. While the husk and the bran are normally already inserted in processes of post-harvest use feeding production processes different from the original one—ranging from the production of energy from biomass to food uses or for breeding—rice straw is generally managed within the same process of cultivation of rice where it is disposed of mainly through two techniques: the burning of straw in the field (also called "burning" or "free burning") and the burying of straw in cultivation land with the aim of returning to the soil part of the nutrients that have been absorbed during plant growth.

The impact on the environment of rice cultivation and the subsequent management of generated wastes is not negligible, particularly with regard to the production of greenhouse gases ("greenhouse gases" GHG). The data available worldwide (Table 7.2) show a contribution of rice cultivation to the production of greenhouse gases from agriculture equal to about 10% of the total and with significant growth forecasts for the coming years: "compared to the average levels of the years 2000, global emissions from rice cultivation are expected to grow by 7 and 6% respectively in 2030 and 2050" [13]. These reflections are even more evident if we consider that by 2050 agriculture will have to grow significantly to contribute to producing food for a world population of 9 billion people [14].

As part of agricultural activities, the cultivation of rice contributes in different ways to the emissions of greenhouse gases: first of all there are GHG emissions, in particular CH_4 natural gas, directly from the rice fields for anaerobic digestion processes, due to the flooding of the rice paddies, of the organic substances contained

Table 7.2 Trend over time (1961–2015) of global emissions of greenhouse gases in the agricultural sector (values in Mton CO_2 eq/year)

Agriculture category	1961	1990	2000	2005	2010	2015
Enteric fermentation	1375	1875	1863	1947	2018	2046
Manure left on pasture	386	578	682	731	764	837
Rice cultivation	366	466	490	493	499	513
Manure management	284	319	348	348	353	345
Crop residues	66	124	129	142	151	211
Total agriculture	2751	4611	4805	4874	5088	5259

Source FAO/FAOSTAT

in the soil that often consist of the same rice straw that is buried as soil nourishment; secondly there are emissions due to the management of production waste that may be due to the combustion of husk in biomass plants for energy production or for free combustion of straw (free burning) aimed at its disposal. The important emissions related to rice straw management processes, whether they are free combustion or burial, are due to the high carbon content of rice straw which is equal to about 40% by mass [15] and which is released into the atmosphere both through combustion processes and through decomposition in an anaerobic environment when the straw is buried and the rice fields are flooded. A further impact generated on the atmosphere by the management of rice wastes, specifically from the practice of free combustion of straw, concerns the emission of pollutants dangerous to human health. The combustion of rice straw generates various pollutants, including particulates (Particulate Matter PM10 and PM 2.5) and polycyclic aromatic hydrocarbons (PHAs "Polycyclic Aromatic Hydrocarbons") [16]. The presence of these components in emissions due to the free combustion of rice straw contributes to increasing the potential toxicity of these emissions [17].

The environmental impacts generated by the free burning of rice straw—often used by farmers as a method of disposing of straw as it is particularly economical—have led over time to the issuing of directives and regulations to regulate and limit these activities. The state of California was one of the first promoters of a burn reduction policy through the publication in 1991 of the "Connelly-Areias-Chandler. Rice Straw Burning Reduction Act" which allows burning only under specific conditions and for disease control. In Europe the regulation (EU)1306/2013 has suggested to Member States to implement a policy of prohibition of field burning of cultural residues. On the subject, shared and validated data on the percentage of rice straw that in Europe is now disposed of through field burning are not yet available; some authors estimate that 30% of the straw produced annually is disposed of by field burning [18].

Considering the emission values (Table 7.3) proposed by the literature [19] it is possible to estimate the impact per unit mass of rice straw that is disposed of through free burning.

To estimate the total amount of straw produced annually in Europe, reference can be made to literature values such as those that estimate production in a range that varies from 3 to 5 tons per hectare [20]; considering the 665,444 ha cultivated with rice in Europe (FAOSTAT data for the year 2016) it is therefore possible to estimate a production variable from 1.99 to 3.32 million tons of rice straw per year of which a quantity between 570,000 and 998,000 tons is estimates that it is burned in the fields—with the unitary emissions referred to in the previous table—and the remaining buried with emissions that are generated after the burial. Methane (CH_4) emissions from the rice fields are directly linked to the practice of incorporating rice straw into the soil because rice fields are located mainly in the anaerobic environment caused by the flooding; this environment foster the activities of the methanogenic bacteria. Different studies also showed that the incorporation of the straw into soil represented the largest contribution to CH_4 emissions, which increased by a variable amount from 65.7 to 79.5% [21–23]. Consequently, although the effects may vary according to the flooding methods of the rice fields, it can be considered that a reduction in the practice

Table 7.3 Examples of emission factors (EFs) specific to rice straw open burning per kg of dry mass [19]

Pollutant	Unit	EF (free burning)	Notes
CO_2	g/kgdm	1460,00	This EF is specific to open field burning of rice straw
CH_4	g/kgdm	1,2	This EF is sourced from the AP-42 database developed by US-EPA for rice straw
N_2O	g/kgdm	0,07	N_2O is one of the gases being emitted from biomass burning and from open field burning of rice straw. A specific EF for rice straw is not available, therefore an average EF for agricultural residues burning is indicated.
CO	g/kgdm	34,7	This EF is specific for open field burning of rice straw
PM2,5	g/kgdm	12,95	This EF is specific for open field burning of rice straw
PM10	g/kgdm	3,7	This EF is specific for open field burning of rice straw
PAH	mg/kgdm	18,62	The EF related to polycyclic aromatic hydrocarbons (PAH) is specific for open field burning of rice straw

of incorporating rice straw into the soil can positively contribute to the reduction of greenhouse gas emissions associated with rice cultivation [24, 25].

With reference to the described scenario, in which straw substantially represents a waste to be disposed of with potentially significant effects on the environment, different research perspectives are currently being opened with the aim of identifying alternative uses for this by-product of rice cultivation. The lines of research that seem most promising to trigger a process of circular economy on a relevant scale are currently the following:

- energy production from biomass: energy production from biomass represented by rice straw is not pursued on a large scale due to the chemical characteristics of the straw that make its combustion products (particularly ashes) particularly aggressive for the plants and difficult to manage. For this reason, research on the production of energy from rice straw is moving towards biomass gasification through anaerobic digestion processes to obtain a biofuel that is easier to manage in production plants or towards the development of straw saccharification processes for the production of alcohol-based fuels [26];
- production of substances for the chemical industry; an example of this research line is the experimentation with the production of levulinic acid from rice straw (e.g. WALEWA Project LIFE13/ENV/ES/001165). Levulinic acid is an intermediate product that is used in various subsequent processes such as the production of biopolymers, plasticizers, biofuels, fertilizers and pesticides and products for the pharmaceutical and cosmetic industries;

- water treatment and purification based on the use of biomass made from rice straw [27] or its derivatives such as coals or ashes;
- treatment of rice straw for the nutrition of animals, offsetting its characteristics that make it different from other straws as it is less rich in lignin and very rich in silica [28];
- production of materials and components for the construction industry; the first attempts to produce building materials starting from rice straw were developed around 1920 [29, 9], compressing rice straw at high temperature or tying it with organic resins. With the process (called "stramit") it was possible to obtain products in the form of panels with variable densities and able to cover several uses ranging from the realization of internal partitions to the production of panels for thermal insulation. Over time, the production of rice straw panels—which in the UK has led to the construction of many thousands of homes—has been decreasing and currently [30] the experiments focus mainly on the creation of materials for thermal insulation. The destination of rice straw to material for thermal insulation can be assessed overall by adding to the benefits obtained by avoiding the free combustion of rice straw (general benefits valid for all circular economy processes that can be activated by rice straw) the advantages related to the consumption of raw material avoided for the production of alternative thermal insulators and the advantages related to the reduction of primary energy consumption for the heating of buildings and the reduction of the consequent emissions of CO_2 into the atmosphere.

7.3.2 Manure Management and Biogas Production

Anaerobic digestion of livestock waste associated with biogas production can be considered as a traditional and consolidated process to produce energy from agricultural scraps and manure. Although some scholars report the use of biogas since the Assyrians, documented production of biogas with anaerobic digesters is traced since mid-nineteenth century when plants were built in New Zealand and India [31]. Nowadays that GHG emissions from livestock are increasing in a significant way (Fig. 4), anaerobic digestion could help reducing methane emissions in agriculture contributing, in the same time, to the generation of renewable energy non subject to discontinuous production [32].

In the last decade there's been a significant increase in the use of biogas technology in the world and especially in Europe where the biogas plants have grown from around 6,000 to around 18,000 in less than 10 years. A little less than 18,000 biogas plants are operating in Europe with a power generation capability of around 10,000 MW compared with the global worldwide power generation capability of around 16,000 MW. The major part of the European anaerobic digestion plants is located in Germany (around 11,000 plants) and in Italy (around 1,600 plants) [33].

The sustainable operation of anaerobic digestion biogas plants should be assessed considering the management of waste and scraps, i.e. the residues (called "digestate") remaining after the anaerobic digestion of biodegradable feedstock. It is estimated that in Europe around 95% of the anaerobic digestion residue is used as an organic fertilizer on agricultural land replacing the use of traditional chemical fertilizers. The use of liquid and solid manure for biogas production is becoming popular in many EU countries and the digestate is very often used as fertilizer and soil amendments. Under certain conditions the digestion residues cannot be used as a fertilizer in an unprocessed form especially in regions that are characterized by an overproduction of animal manure in comparison to the available arable land [34]. The high density, in certain regions, of the anaerobic digestion plants and the consequent limited arable land available for using the process residue may lead to overcharge fields with digestate; the digestate overcharge cause an excessive nitrogen load on soil that may have negative effects on soil properties and underground water quality [35]. The strong increase in the number of biogas plants could therefore result in an imbalance between the availability of digestion by-products (digestate) and the availability of agricultural land on which to use such by-products as fertilizers (as is normally the case now). The use of the digestion residues as fertilizer shall be done under the regulatory constraints on the use of nitrogen compounds in agriculture. The European Nitrates Directive (91/676/EEC) promotes the rationalization of the use in agriculture of manure and similar effluents by providing that these distributed fertilizers do not exceed the needs of crops. The Nitrates Directive involves the fact that the nitrogen supply to agricultural land shall be limited to a quantity not exceeding 170 kg of nitrogen per hectare per year (a value that can be raised up to 250 kg of nitrogen per hectare per year in specific areas which must be identified at regional level). This means that the reuse of the anaerobic digestion residues, particularly rich in nitrogen, may have a limit which is determined case by case through local regulations. If it is not possible to completely re-use the digestate, this should be used in areas other than its production, resulting in significant transport costs for a low-value product.

It would be useful to identify alternative forms of use for the solid digestate that, in addition to reducing the environmental impacts associated with its transport to other destinations, may allow to create added value to the digestion by-product incorporating it into manufactured products for construction industry or for other designed products. It is believed that the use of the digestate as soil improver and fertilizer is interesting (and makes the biogas plant economically convenient) if it is possible to limit the transport distances of the digestate to the land of use within 10–20 km [36]. The literature review shows that the residues of the anaerobic digestion plants are re-used within the agricultural industry with different limitations as the low added value and the impact on the environment arising from the transport of a large quantity of liquid digestate and solid digestate [37–39]. According to this, it can be interesting and even useful to increase the diffusion of anaerobic digestion plants to find alternative uses for these residual solids coming from anaerobic digestion plants [40] and the studies on the use of dry solid digestate from anaerobic digestion plants show that opportunities are present for its use in composite materials [41].

Starting from these considerations, researchers are trying to assess the suitability of the digestate to be used in other fields than the agriculture especially for what it concerns the solid part of the digestate that includes a wet lignocellulosic based

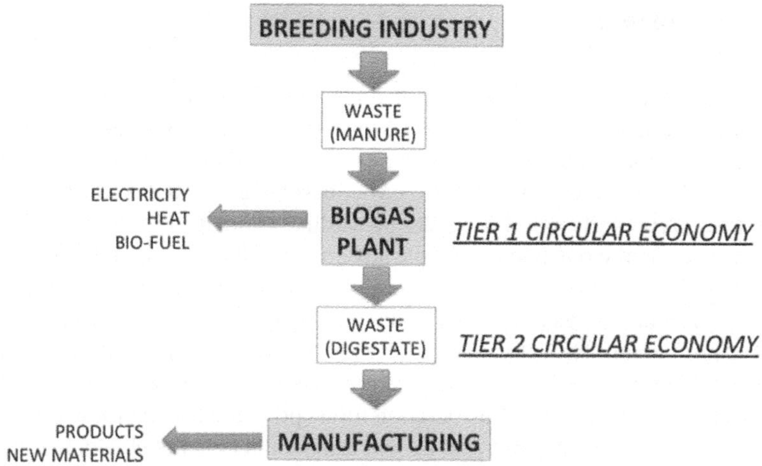

Fig. 7.6 Second tier circular economy approach

fibrous residue. The lignocellulosic residuals embodied in the solid part of the anaerobic digestion residues can be used in combination with wood for the making of engineered wood composites such as hardboard, particleboard, or Medium Density Fiberboard (MDF) or bio-based thermoplastic composites [42]. In literature, no or little studies on digestate-composites based resin have been found [43]. Some studies have been found on the use of dry solid digestate in the manufacturing of particle wood board [40]. An Italian experimentation of the use of digestate for production of composite materials is reported by the press as an attempt to "close the loop" in the bovine manure management [44]. Some studies have been found on an alternative use of dry solid digestate which was addressed to supply pyrolysis thus increasing the net energy gains and obtaining "biochar" to be used as soil amendment alternatively to solid-digestate [35]. In Germany and Austria, part of the fibers coming from the solid fraction of digestate are used for the production of biomaterials and biocomposites for the automotive and construction industries [45]. The above mentioned studies fall within an approach oriented to improve sustainability of the manure management processes in the livestock industry with the application of a circular economy strategy based on a two tiers approach (Fig. 7.6): first tier is the reuse of the waste of the main process (manure) to produce biogas; second tier is the reuse of the waste of the tier 1 process (the digestate) to produce industrial or handicraft products with high added value.

7.4 Conclusions

The agriculture and breeding sector is in constant and continuous expansion due to the growing demand for food: this growth has as a direct consequence the increase of waste generation—normally problematic to dispose of or reused with a low added value—and the increase in greenhouse gas emissions generated from this sector. It is therefore particularly interesting the possibility of activating circular economy processes that, on the one hand, can take advantage of the large quantity of material made available in the form of sector waste and, at the same time, may allow to reduce the sector's greenhouse gas emissions as has already been done in many other areas of industrial production. The activation of these processes of circular economy should be evaluated taking into account: the characteristics and the quantitative aspects related to the waste that is taken into consideration for its potential use in production chains, even different from the original one; the environmental benefits obtainable from the activation of these processes. Starting from the definition of a reference baseline, it is possible to identify the environmental aspects and estimate the positive environmental impacts in terms of reduction of the environmental loads generated by the waste disposal processes produced by the analysed industrial, agricultural or extractive processes. Such analyses can make these virtuous mechanisms of material recovery from the agricultural sector even more interesting when a proven and consistent environmental benefit is added to the economic benefit. The analysed cases, relating to the rice straw and the manure management, show that a global assessment of the environmental loads of the upstream supply chain (waste production) and the downstream supply chain (use of waste) is necessary to evaluate the opportunities for developing circular economy processes that should always be evaluated in their entirety for a better understanding of the benefits associated with them.

References

1. United Nations (1999) The world at six billion
2. FAO (2018) World food and agriculture—statistical pocketbook 2018. Rome
3. IPCC (2014) Climate Change 2014: mitigation of climate change. contribution of working group III to the fifth assessment report of the intergovernmental panel on climate change, Cambridge University Press, Cambridge, United Kingdom and New York, NY, USA
4. Berna F (2017) Geo-ethnoarchaeology study of the traditional Tswana dung floor from the Moffat Mission Church, Kuruman, North Cape Province, South Africa. Archaeol Anthropol Sci 9(6):1115–1123
5. Anderson S, Ertug-Yaras F (1998) Fuel fodder and faeces: an ethnographic and botanical study of dung fuel use in central Anatolia. Environ Archaeol 1(1):99–109
6. Agamuthu P (2009) Challenges and opportunities in agro-waste management: an Asian perspective. Inaugural meeting of First Regional 3R Forum in Asia, Tokyo, Japan, 11–12 Nov 2009
7. El-Haggar SM (2007) Chapter 7—sustainability of agricultural and rural waste management. In Sustainable industrial design and waste management, Academic Press, pp 223–260
8. EN 16575:2014 Bio-based products—vocabulary

9. Madurwar MV, Ralegaonkar RV, Mandavgane SA (2013) Application of agro-waste for sustainable construction materials: a review. Constr Build Mater 38(2013):872–878
10. Mohanty A, Vivekanandhan S, Pin JM, Misra M (2018) Composites from renewable and sustainable resources: challenges and innovations. Science 362:536–542
11. Pappua A, Saxenaa M, Asolekarb SR (2007) Solid wastes generation in India and their recycling potential in building materials. Build Environ 42:2311–2320
12. Raut SP, Ralegaonkar RV, Mandavgane SA (2011) Development of sustainable construction material using industrial and agricultural solid waste: a review of waste-create bricks. Constr Build Mater 25(2011):4037–4042
13. FAO Agriculture (2014) Forestry and other land use emissions by sources and removals by sinks, ESS Working Paper no. 2
14. COM 112(2011) E/COM, a roadmap for moving to a competitive low carbon economy in 2050
15. Air Resources Board—California Environmental Protection Agency—Research division (1997) Alternative uses of rice straw in California, San Luis Obispo, CA, USA
16. Estrellan CR, Lino F (2010) Toxic emissions from open burning. Chemosphere 80:193–207
17. Oanh NTK, Thuy Ly B, Tipayaroma D, Manandhar B, Prapat P, Simpson CD, Liu L (2011) Characterization of particulate matter emission from open burning of rice straw. Atmos Environ 45:493–502
18. Moliner C, Bosio B, Arato E, Ribes A (2014) Comparative study for the energy valorization of rice straw. Chem Eng Trans 37:241–246
19. Gadde B, Bonnet B, Menke C, Garivait S (2009) Air pollutant emissions from rice straw open field burning in India. Thailand Philippines, Environ Pollut 157:1554–1558
20. Chinese D, Monaco B, Nardin G, Simeoni P (2006) Energia dalle biomasse—Le tecnologie, i vantaggi per i processi produttivi, i valori economici e ambientali, Rapporto finan-ziato da UE Obiettivo 2 FSER
21. Devevre O, Horwàth W (2000) Decomposition of rice straw and microbial carbon use efficiency under different soil temperatures and moistures. Soil Biol Biochem 32:1773–1785
22. Ma J, Xu H, Yagi K, Cai Z (2008) Methane emission from paddy soils as affected by wheat straw returning mode. Plant Soil 313:167–174
23. Xionghui J, Jiamei W, Hua P, Lihong S, Zhenhua Z, Zhaobing L, Faxiang T, Liangjiec H, Jianc J (2012) The effect of rice straw incorporation into paddy soil on carbon sequestration and emissions in the double cropping rice system. J Sci Food Agric 92:1038–1045
24. Miura Y, Kanno T (1997) Emissions of trace gases (CO_2, CO, CH_4 and N_2O) resulting from rice straw burning. Soil Sci Plant Nutr 43:849–854
25. Neue U, Wassmann R, Lantin S, Alberto R, Aduna JB, Javellana M (1996) Factors affecting methane emission from rice fields. Atmos Environ 30(10/11):1751–1754
26. Ranjana A, Moholkara VS (2013) Comparative study of various pretreatment techniques for rice straw saccharification for the production of alcoholic biofuels. Fuel 112:567–571
27. Amin MN, Mustafa AI, Khalil MI, Rahman M, Nahid I (2012) Adsorption of phenol onto rice straw biowaste for water purification. Clean Techn Environ Policy 14:837–844
28. Van Soest PJ (2006) Rice straw, the role of silica and treatments to improve quality. Animal Feed Sci Technol 130(3/4):137–171
29. Yates T (2006) The use of non-food crops in the UK construction industry. J Sci Food Agric 86:1790–1796
30. Wei K, Lv C, Chen M, Zhou X, Dai Z (2015) Development and performance evaluation of a new thermal insulation material from rice straw using high frequency hot-pressing. Energy Build 87:116–122
31. Bond T, Templeton MR (2011) History and future of domestic biogas plants in the developing world. Energy Sustain Dev 15(2011):347–354
32. Lyng K, Bjerkestrand M, Elstad Stensgård A, Callewaert P, Jørgen Hanssen O (2018) Optimising anaerobic digestion of manure resources at a regional level. Sustainability 10:286. https://doi.org/10.3390/su10010286
33. EBA (2014) European biogas association: number of biogas plants in Europe. http://european-biogas.eu/wp-content/uploads/2015/02/Biogas-graph-20131.png. Accessed 04.04.18

34. Vaneeckhaute C, Meers E, Michels E, Buysse J, Tack F (2013) Ecological and economic benefits of the application of bio-based mineral fertilizers in modern agriculture. Biomass Bio-energy 49:239–248. https://doi.org/10.1016/j.biombioe.2012.12.036
35. Monlau F, Francavilla M, Sambusiti C, Antoniou N, Solhy A, Libutti A, Zabaniotou A, Barakat A, Monteleone M (2016) Toward a functional integration of anaerobic digestion and pyrolysis for a sustainable resource management, Comparison Between Solid-Digestate and Its Derived Pyrochar as Soil Amendment. Appl Energy 169(2016):652–662
36. Fabbri C, Piccinini S (2012) Bovini da latte e biogas. Linee guida per la costruzione e la gestione di impianti. Centro Ricerche Produzioni Animali, Reggio Emilia
37. Alrefai R, Benyounis KY, Stokes J (2017) Integration approach of anaerobic digestion and fermentation process towards producing biogas and bioethanol with zero waste: technical. J Fundam Renew Energy Appl 7:6. https://doi.org/10.4172/2090-4541.1000243
38. Divya D, Gopinath LR, Merlin Christy P (2015) A review on current aspects and diverse prospects for enhancing biogas production in sustainable means. Renew Sustain Energy Rev 42(2015):690–699
39. Kataki S, Hazarika S, Baruah DC (2017) Assessment of by-products of bioenergy systems (anaerobic digestion and gasification) as potential crop nutrient. Waste Manag 59(2017):102–117
40. Spelter H, Winandy J, Zauche T (2008) Anaerobically digested bovine biofiber as a source of fiber for particleboard manufacturing: an economic analysis. BioResources 3(4):1256–1266
41. Matuana L, Gould MC (2006) Promoting the use of digestate from anaerobic digesters in composite materials. Final Report: Grant #PLA-06-42. Community Energy Project. Michigan State University, East Lansing, MI
42. Winandy JE, Cai Z (2008) Use of ADBF for Fiberboard. BioResources 3(4):1244–1255
43. Taurino R, Lancellotti I, Tatàno F, Carchesio M, Pozzi P (2016) Mechanical and chemical resistance of composite materials with addition of anaerobic digestate. Composites Part B 92:e259–e264
44. Smalwood C (2016) Cow dung goes high design. The New York Times Style Magazine
45. Consorzio Italiano Biogas (2014) Biogas 2020. 25.000 green jobs per l'agricoltura italiana. Terzo position paper del biogas Italiano

Chapter 8
Information and Communication Technologies (ICTs) for Advanced Scraps/Waste Management

Abstract The present chapter introduces the topic of Smart Waste Management enabled by the adoption of innovations in the field of Information and Communication Technology (ICT). The complex nature of the waste management challenge requires coordination and collaboration, pooling of resources and support for the definition of global goals and strategies. In order to guarantee the achievement of these objectives, it is fundamental to recognize the role of technological innovations in the ICT field as useful tools to establish active and collaborative cross-sectoral networks aimed at identifying, evaluating and implementing virtuous processes of industrial symbiosis. In particular, the concept of Information Platform is introduced as a potential interoperable framework aimed at sharing, integrating and exchanging information between stakeholders at different territorial scales, with a view to the centralized optimization of information. The approaches to the topic of Smart Waste Management and the ICT tools adopted by the United Nations Environment Programme (UNEP) and the European Union (EU) are also analyzed and discussed to outline present and future interests and trends. Lastly, the chapter analyzes virtuous Smart Waste Management projects funded by the European Commission under the Horizon 2020 Programme.

Keywords Advanced waste management · Smart waste management · Information and communication technology (ICT) · Symbiotic systems · Big data · Information platform · Information sharing · Information management

8.1 Symbiotic Systems and ICT-Based Waste Management

In recent years, the concept and the approach to waste are undergoing a profound transformation also due to the introduction of increasingly ambitious targets and management strategies aimed at achieving more sustainable scenarios. According to

This chapter is authored by Nazly Atta, Politecnico di Milano, Department of Architecture, Built Environment and Construction Engineering, Milan, Italy. e-mail: nazly.atta@polimi.it.

M. Migliore et al., *Strategies for Circular Economy and Cross-sectoral Exchanges for Sustainable Building Products*, Springer Tracts in Civil Engineering,
https://doi.org/10.1007/978-3-030-30318-1_8

the United Nations Environment Programme (UNEP)[1] report of 2011: "Towards a Green Economy. Pathways to Sustainable Development and Poverty Eradication", the recycling targets for 2050, with a view to green economy, provide recycling percentages of 15% for industrial waste, 34% for urban waste and 100% for electronic waste [1, 2]. Furthermore, in the same report, UNEP stresses that the waste recycling sector is one of the most important in terms of employment development potential. The estimates of the report show, in fact, that the recycling sector creates ten times more jobs than the disposal and incineration sectors.

Although the path is still very long, in the last years the approach to the issue of waste has radically changed both as regards the private sphere of the individual citizen and companies and the public sphere, thanks to the introduction of waste management policies aimed at achieving goals with increasing efficiency.

Within this transforming scenario, the efficient management of waste-related data and information plays a fundamental role.

The waste seen as a source of resources shared and exploited on a territorial scale introduces profound reflections on the potential of an active and collaborative network capable of enabling virtuous inter-sector circles of "waste-resource" aiming to achieve the objectives of a circular economy [3].

Analyzing the concept of circular economy, already introduced in the previous chapters of this book, it is possible to observe how it is closely linked to the concept of industrial symbiosis.[2] Indeed, the industrial symbiosis can be defined as a form of intermediation, or a policy tool, designed to encourage and facilitate an innovative collaboration between companies, also belonging to different business sectors, in such a way that the waste produced by one of them is valued as raw materials for the other.

Thus, the models of industrial symbiosis involve exchanges of flows of both materials and energy, in the context of more or less concentrated industrial complexes, in order to obtain a virtuous industrial ecosystem, able to optimize the consumption of energy and materials, which uses waste of a process as the production input for other processes [4].

In this way, in a process of industrial symbiosis, the different stakeholders—belonging to different business sectors—can derive mutual benefits from the sharing of waste by transforming the waste of one sector into raw material useful for the production processes of another sector.

---

[1]The United Nations Environment Programme (UNEP) is the "*leading global environmental authority that sets the global environmental agenda, promotes the coherent implementation of the environmental dimension of sustainable development within the United Nations system, and serves as an authoritative advocate for the global environment*". Source: UNEP website, www.unenvironment.org.

The main mission of UNEP is "*to provide leadership and encourage partnership in caring for the environment by inspiring, informing, and enabling nations and peoples to improve their quality of life without compromising that of future generations*". Source: UNEP website, www.unenvironment.org.

[2]According to Costa et al. [4], the industrial symbiosis "*emerged as a self-organizing business strategy among firms that are willing to cooperate to improve their economic and environmental performance*" [4].

In order to achieve this valuable goal, it is necessary to implement a significant systemic innovation among stakeholders on a territorial scale. Indeed, industrial symbiosis requires a broad coordination between the various stakeholders (such as industry, public authorities and policy makers, academia and research, civil society organizations, etc.) and a greater awareness by industries—which have a key role in economic and social transformation—of the responsibilities related with the production of industrial waste [5].

In order to effectively develop symbiotic systems, it is necessary to consider that the complexity and heterogeneity of waste flows require coordination and networking among stakeholders (e.g. private companies and entrepreneurs, researchers, public authorities, etc.) for:

- harmonizing technologies, processes and services;
- profiting from benchmarking opportunities;
- sharing best practices;
- using or/and developing industry standards.

In fact, an insufficient cooperation along and among the value-chain of different sectors represents an obstacle to the achievement of the symbiotic system. In particular, the poor cooperation and collaboration of value-chain can be translated into scenarios characterized by a low efficiency if analyzed from an environmental and socio-economic point of view, both in terms of lower recycling rates and of suboptimal use of raw materials.

Moreover, a primary challenge in achieving the objective of industrial symbiosis—with a view to circular economy—is the one linked to the ability to collect, access and share information—such as sources, quantity, quality of waste streams and resources required—especially in the complex reality of large cities [3].

The global nature of the waste management challenge if on the one hand holds the potential to export eco-innovative solutions and to seize new markets, on the other it requires: coordination, collaboration and sharing of resources among the involved actors, as well as support tools useful for the definition of common goals and strategies. In particular, in this regard the international dissemination of knowledge on waste management—including reference environmental regulations and standards—can support the transformation of waste into a global resource and the establishment of efficient systems, tools, technologies and services for an advanced waste management, particularly in developing countries and emerging economies.

Therefore, advanced forms of participatory processes that involve all the stakeholders are necessary. Furthermore, it is necessary to develop a strategy for the dissemination and the adoption—on a global scale—of best practices regarding integrated waste management and for the sharing of reference standards and benchmarks.

In order to guarantee the achievement of these objectives, it is of fundamental importance to recognize the role of technological innovations in the field of Information and Communication Technologies (ICTs). In fact, nowadays ICTs allow the creation and the development of interoperable, integrated and shared knowledge bases with a view to centralized information optimization (e.g. information platform as a support for the capitalization and sharing of information).

The presence of a central unique dynamic database—built according to unified data collection and processing protocols—able to collect and store data and information concerning materials or components throughout their whole life cycle—including also real-time data coming from advanced real-time monitoring systems—becomes essential to strengthen decision-making processes and capabilities at local, regional and national levels. Indeed, the presence of a real-time, integrated and interoperable monitoring system able to collect (and send to the database) feedback data and information coming from the management phase of material, components and scraps/waste (e.g. information about waste quantities, status of waste collection and treatment, etc.) represents an element of crucial utility, to be added to historical and statistical information.

In particular, such real-time monitoring system is connected to the central database and it consists of a disseminated network of devices and sensors of the Internet of Things (IoT),[3] able to acquire data and information through innovative methods and technologies. Such devices, connected to the Internet, allow the collection of data—in the characteristics of Big Data—and their transmission through cloud networks for their storage and subsequent processing. Big Data can be defined as a complex set of data characterized by [6–8]:

- high Volume. Large data whose orders of magnitude can reach the exabytes (EB) and zettabytes (ZB);
- high Velocity. High speed—up to the real time—of generation, collection, updating, processing and analysis of data flows in and out of the network systems;
- high Variety. Multiple data sources (e.g.: texts, documents, numerical data, geospatial data, audio, photos, videos, etc.) and data heterogeneity (e.g.: structured, semi-structured, unstructured data, etc.);
- high Variability. Variability in terms of data meaning (meaning of data can change according to the change of user or the examined aspects) and in terms of data lifecycle (data can change their value and their reliability over time, some data may become obsolete or no longer valid);
- high Veridicity. The level of accuracy and reliability of the data become less and less measurable, controllable and manageable due to the heterogeneity of informal sources such as, for example, data and information on social networks or on the web in general;
- high Value. The value lies in the high information potential. The construction of an ever wider knowledge base increases decision-making skills, offering the possibility of answering questions that previously were impossible to answer.

[3]The Internet of Things (IoT) is defined by the International Telecommunication Union—Telecommunication Standardization Sector (ITU-T) as "*a global infrastructure for the information society, enabling advanced services by interconnecting (physical and virtual) things based on existing and evolving interoperable information and communication technologies*". Source: ITU-T website, www.itu.int.

Moreover, Gartner defines IoT as "*the network of physical objects that contain embedded technology to communicate and sense or interact with their internal states or the external environment*". Source: Gartner IT Glossary website, www.gartner.com/it-glossary.

In this regard, this data complexity can be properly managed through an information platform defined as an open, interoperable and structured digital system capable of managing the heterogeneity of data deriving from different sources in a single central database, as well as the simultaneous presence of multiple stakeholders interested in the information bases and digital services offered by the platform itself [9].

Therefore, an information platform can represent an interoperable framework that aims to promote, share and exchange information on issues related to waste among the various stakeholders, at different territorial scales. In this approach, this dynamic information tool is able to provide the information necessary for the identification and evaluation of possible processes of industrial symbiosis.

In fact, the implementation of a platform allows the various stakeholders to interact, share information and make their contribution mainly in relation to:

- the definition of an integrated research and strategic innovation program;
- the definition of virtuous eco-innovation models and circular and systemic business models aimed at the prevention and efficient management of waste;
- the strengthening of the bonds and synergies between the different sectors, in order to consolidate value-chains, identify and evaluate mechanisms of industrial symbiosis, and develop new market opportunities.

An important step forward for companies towards new virtuous and efficient forms of waste management, with a view to industrial symbiosis, is represented by the knowledge of their own waste. The process of acquisition and validation of data made available and shared by the various stakeholders interacting with a possible platform is a crucial goal for the sector innovation.

Therefore, the adoption and implementation of an Information Platform can support stakeholders in reaching this objective of cooperation and collaboration and information sharing. In particular, the digital platform acts as an open system able to gather information, coming from various sources, into a unique shared database. The collected information may concern: typology, features and quantities of waste materials and products, as well as the related management processes.

In this way, it is possible to increase the knowledge necessary to optimize decision-making processes and, above all, to allow circular and virtuous processes of industrial symbiosis, since each stakeholder is informed about:

- its own scraps/waste, offered to other stakeholders;
- the waste of the other involved stakeholders, acquirable as a resource for its own production processes.

An important task in developing a platform is to analyze the available data. The current information platforms, in fact, allow to carry out structured processes of inspection, cleaning, transformation, modeling, interpretation and visualization of data with the aim of discovering useful information and insights, thus informing and supporting decision-making processes.

Moreover, a platform is able to let the users visualize data, information and indicators—through graphics, diagrams and other user-friendly representations easy to comprehend—useful to develop strategies for:

- optimizing operational performance;
- reducing processing costs;
- increasing incoming economic flows;
- identifying new business opportunities, laying the foundations for a true circular economy.

In particular, by analyzing the potential of a platform, the following four main functions can be highlighted.

8.1.1 Data Collection

Through an information platform it could be possible to automatically import a wide range of waste management data in a normalized and standardized way according to a proper pre-defined terminology and hierarchical structure. For example, it is possible to enter data and information such as:

- quantity of scraps/waste;
- type of scraps/waste;
- address, date and time of collection;
- frequency and details of the waste collection routes, customer locations and location of containers/waste collectors;
- general and detailed information about waste management companies;
- waste transporter vehicles, details including dimensions of the truck/s, truck typology, identification plates, license number of the driver/s, etc.;
- weight scale for both the transport vehicle and waste containers;
- financial data, such as transportation fees/commissions, value of the transported material, etc.;
- general and detailed data non properly referred to waste but to the context conditions, such as climatic conditions, population density, traffic data, etc.

Moreover, it is important to stress that Information Platforms are expandable open systems, thus it is possible to start with an information base made up by the available data and to continue overtime to expand the database with further information as well as to increase the level of detail of information according to the users' needs.

8.1.2 Data Standardization and Normalization

The heterogeneous nature[4] of data (Table 8.1) related to the scraps/waste sector (data from different sources, data in different shapes and formats, data owned by different

[4]The wide variety of IoT data acquisition devices lead to have a richness of data that, however, should be properly rationalized in order to be able to exploit its potential. Data coming from the different IoT devices are heterogeneous, thus they have high variability of data types and formats

Table 8.1 Heterogeneity of data. Adapted from: Wang [10]

Type of data heterogeneity	Description
Syntactic heterogeneity	The different data sources are not expressed in the same language
Conceptual (or semantic) heterogeneity	Differences in the modeling of the same domain of interest
Terminological heterogeneity	Variations in the denomination (and coding) of the same entities by different data sources
Semiotic (or pragmatic) heterogeneity	Different interpretation of entities by different people and involved parties

stakeholders, etc.) implies the need of data standardization and normalization processes in order to be able to exploit the value of the whole central database of the Information Platform.

In particular, through the normalization process it is possible to gather into a unique and centralized database information and data according to specific and shared structures and procedures. This allows to minimize data redundancy, ambiguity and dependency.

In parallel, the process of data standardization aims to provide a same language for data coming from heterogeneous sources and which may have heterogeneous shapes and formats by implementing business rules around abbreviations, synonyms, patterns, casing, or order matching. This process of data "cleaning" enables to avoid redundancies and inconsistencies leading to a better data quality.

---

[10]. They are possibly ambiguous and low quality due to missing values, high data redundancy, and untruthfulness.

Moreover, the IoT devices can cover large-scales and they can constitute a distributed network across the city where each data acquisition device is placed at a specific and determined geographic location. This implies a strong correlation between time and space. This correlation can only be exploited if pre-processing activities—data standardization and normalization—are performed on the massive data coming from the multiplicity of IoT data sources.

Through standardization and normalization processes it is possible to [10]:

- improve the quality and accuracy of data;
- reduce redundancies to improve the database performance;
- facilitate the integration, analysis and comparison with historical databases;
- develop an accurate benchmarking system for assessing present and future performance;
- increase the efficiency in operations;
- improve and optimize decision-making processes that can be based on updated, high-quality and reliable data.

Indeed, standardization and normalization represent the needed data pre-processing to provide a consistent and reliable information base to the next level of data analytics.

8.1.3 Data Sharing

IoT capabilities involve the possibility to share data (also in real-time) gathered by devices and sensors. This possibility represents a novel opportunity for the operators in waste management field since up to now data and information were confined to proprietary systems which made it impossible to share data, exploiting their value and the multiple insights that could arise by detecting correlations or relationships among data themselves.

In particular, Information Platforms allow to overcome the one-to-one relationships to reach one-to-many and many-to-many approaches where the platform has the function of a digital place which connects stakeholders with different interests that are now able to share their "waste data and information" identifying, accessing and exploiting possibilities to establish synergies or to learn from the experiences of the others.

Therefore, the collaboration among stakeholders to jointly develop business opportunities—identified on the basis of sharing and consequent combination of all data segments—represents a fundamental function that nowadays Information Platform are able to guarantee. The platform could be understood as a valuable and comprehensive source of data that all parties could use to improve their knowledge, their strategies and their practices.

8.1.4 Data Analysis

A platform can provide a wide range of data analysis and visualization capabilities in a single environment to help the users to better understand complex data sets by enabling:

- comparisons between actual results and expected results relating to different areas of interest (e.g. waste characterization, financial data, etc.);
- identification of missing documentation or information;
- assessment of the achievement of objectives through the use of KPI (Key Performance Indicators);
- identification of correlations or trends not identifiable by traditional methods of data analysis.

The tools for analyzing and visualizing data can also enable to:

- visualize dashboards that provide managers with a global overview of financial data on waste management, performance indicators, and the variance between achieved and expected objectives;
- display reporting tools, with interactive real-time filtering capabilities to perform dynamic trend analysis.

As regards the different techniques of data analysis, it is possible to mention:

- Descriptive Analytics. Descriptive analyzes allow to access data thanks to flexible logical views and to visualize the main performance indicators in a synthetic and graphic way. In particular, descriptive analytics aim to describe what is happening and to provide to decision-makers the useful information to describe, represent and interpret reality using visualization tools. The descriptive analyzes analyze past events, at different levels of detail depending on the purpose, to understand and outline dynamics and performance of the parameters of interest, finally extracting information and useful indications for appropriately approaching future activities.
- Predictive Analytics. Predictive analytics aim to predict actions that will occur in the near future. They use mathematical techniques (e.g. regression, forecasting, predictive models, etc.) to provide future forecasts and future development scenarios. In particular, predictive analyzes aim to predict the behaviors and performance of the parameters of interest and activities, through data modeling and scenario simulation tools. In this way, in addition to the detection of present trends, it is possible to foresee future behaviors thus supporting decision-making processes at different levels, as well as the development of strategies and the identification of new business opportunities.
- Prescriptive Analytics. Prescriptive analytics can be described as advanced tools that combine data analysis with the ability to take and manage decision-making processes. Prescriptive Analytics provide strategic indications or operational solutions based on both descriptive and predictive analysis. Prescriptive analyzes are in fact capable of proposing strategic and operational opportunities to decision-makers on the basis of the results of data analyzes. In particular, these analyzes exploit simulation and optimization algorithms to advice on possible outcomes, thus allowing decision-makers to identify and assess a number of different possible actions, leading towards a proper solution.

8.2 UNEP Approach: The GPWM Knowledge Platform

The United Nations Environment Programme (UNEP)[5] is the world's leading authority on environmental matters. It defines the global environmental agenda and promotes the implementation of the environmental component of sustainable development within the United Nations system. The activity of UNEP includes:

- the assessment of environmental conditions and trends on global, national and regional levels;

[5]The United Nations Environment Programme (UNEP) represents the main world authority on environmental issues. The mission of UNEP is: "*to provide guidance and encourage collaboration to protect the environment, inspiring, informing and allowing nations and populations to improve their quality of life without compromising that of future generations*". Source: UNEP website, www.unenvironment.org.

– the development of national and international environmental information tools;
– the strengthening of institutions for an environmentally conscious management.

In the area of Waste Management, according to UNEP, the implementation of the so-called Environmentally Sound Technologies (ESTs)[6] is the key to achieve a sustainable waste management that is effective and efficient. The ESTs include, among the technological innovations, those technologies that have the potential to significantly improve environmental performance. According to the UNEP IETC,[7] ESTs are those technologies that are: less polluting, use all resources more sustainably, recycle more than their waste, and manage residual waste in a more acceptable way—from the environmental point of view—than the technologies they replace (UNEP website: www.unenvironment.org).

However, the ESTs are not to be understood only as technological tools or products, but as a global system which also includes know-how, goods and services, procedures, and organizational and management models.

Therefore, in order to disseminate and share this system and make it global, it is necessary to develop information networks that are able to connect international, national and regional realities.

Furthermore, the collection and sharing of information is essential to strengthen decision-making processes at local, regional and national level. For this purpose, the UNEP has created the Global Partnership on Waste Management (GPWM).[8]

The Global Partnership on Waste Management (GPWM) (Fig. 8.1) is a permanent partnership between international organizations, governments, companies, academia, local authorities and non-governmental organizations. This partnership was launched in November 2010 in order to:

---

[6]According to UNEP, the Environmentally Sound Technologies (ESTs) are "*technologies that have the potential for significantly improved environmental performance relative to other technologies. ESTs protect the environment, are less polluting, use resources in a sustainable manner, recycle more of their wastes and products, and handle all residual wastes in a more environmentally acceptable way than the technologies for which they are substitutes. ESTs are not just individual technologies. They can also be defined as total systems that include know-how, procedures, goods and services, and equipment, as well as organizational and managerial procedures for promoting environmental sustainability*". Source: https://www.unenvironment.org/regions/asia-and-pacific/regional-initiatives/supporting-resource-efficiency/environmentally-sound.

[7]The International Environmental Technology Center (IETC) is a division of UNEP. The main function and goal of the UNEP IETC is to promote and facilitate the application of ESTs by offering consulting services and technical support to the governments of developing/in transition countries. Source: UNEP IETC website, www.unenvironment.org/ietc.

[8]According to UNEP, the Global Partnership on Waste Management (GPWM) is "*an open-ended partnership for international agencies, Governments, businesses, academia, local authorities, and nongovernmental organizations. The GPWM supports the development of work plans to facilitate the implementation of integrated waste management at national and local levels to overcome environmental, public health, social and economic problems arising from unsound waste management. The GPWM supports policy dialogues and other activities to exchange experiences and practices, facilitating enhanced awareness raising and capacity building.*" Source: GPWM 2016. Framework Document (available at: www.sids2014.org/content/documents/commitments/9501_7462_commitment_The%20GPWM%20Framework%20Document.pdf. Accessed in April 2019).

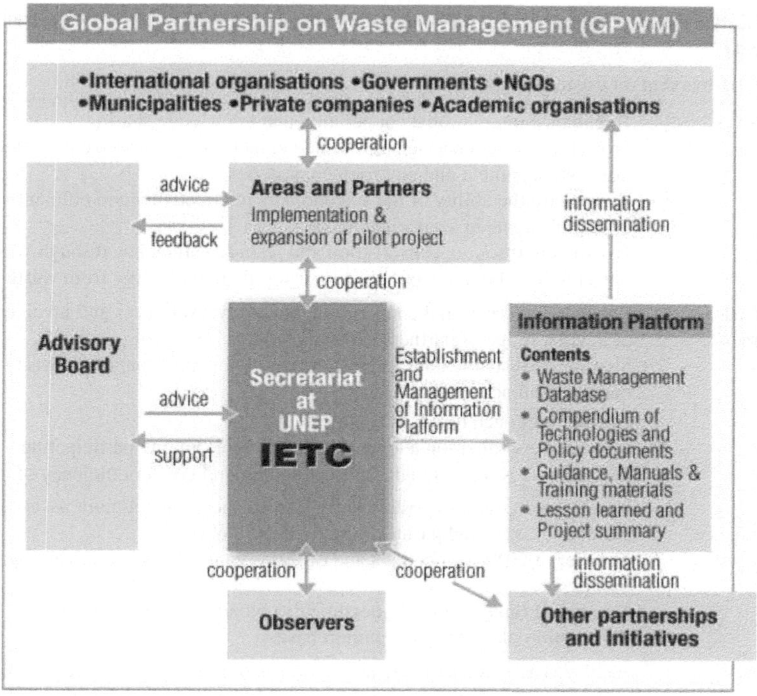

Fig. 8.1 Structure of the Global Partnership on Waste Management (GPWM). *Source* UNEP 2011(GPWM: framework document)

- improve international cooperation among stakeholders;
- identify and fill information gaps;
- share information and strengthen awareness;
- promote resource conservation and efficiency (Table 8.2).

Moreover, the UNEP has made available to the GPWM and to all the interested parties globally, an information platform: the Knowledge Platform. This platform was launched for guiding users to existing databases on waste management, facilitating in this way information sharing and coordination among all the involved stakeholders (UNEP 2016).

In particular, the GPWM Knowledge Platform (Fig. 7.1) is an online portal that aims to promote, share and exchange information about Waste Management.

The GPWM Knowledge Platform aims to:

- guarantee, in particular to developing countries, the access to scientific and technological information as well as to information and reports concerning the state of the art in waste management;
- promote and facilitate the transfer of know-how;
- enhance the capabilities of the stakeholders in assessing, adopting and implementing the ESTs;

Table 8.2 General information about GPWM. Adapted from: GPWM 2016 (GPWM: framework document)

Global partnership on waste management	
Objectives	– To enhance international cooperation and knowledge sharing – To create win-win partnerships aimed at increasing the awareness about waste management and environmental issues – To increase the ability of the stakeholders in effectively and efficiently managing waste in a sustainable way – To promote resource conservation and resource efficiency through waste prevention and to recover valuable material and/or energy from waste
Expected outcomes	– Enhanced international cooperation, advocacy, awareness and involvement of a wider range of partners and stakeholders – Sharing of available data and generation of new information on waste management activities – Holistic approach to waste management – Improved coordination and increased synergies among participating institutions, as well as better use of funding and greater efficiency of efforts
Key performance indicators	– Number of activities carried out by lead partners on biennium basis – Number of new lead partners joining the GPWM – Number of GPWM meetings and events organized collaboratively by lead partners – Number of links within the documents uploaded into the GPWM information platform on biennium basis
GPWM strategy	The GPWM acts as a *forum* to share information and improve the coordination of activities carried out by the various engaged stakeholders, enabling them to: – complement each other's efforts – reduce duplication – create synergies – optimize the efficiency of operations and resources – promote collaboration The GPWM aims to promote coherence in, and coordination of, actions on waste management at the global scale, towards a global and comprehensive effort
GPWM added value	– Sharing information and exchanging knowledge and best practices to help empower policymakers and practitioners to undertake well-informed decision-making and efficient implementation of policies and programmes on waste management – Building a multi-stakeholder network of experts in waste management by: facilitating working groups, coordinating activities, building partnerships and synergies among stakeholder, bridging in this way the gaps in current efforts with the aim to improve efficiency and effectiveness – Pooling resources by centralizing the request for and the coordination of: external financing, technical resources and assistance, maximizing in this way the cost effective use of scarce resources, thus contributing to support at all levels the identified challenges of waste management

– strengthen the collaboration among the engaged stakeholders.

As regards the structure, the GPWM Knowledge Platform has a modular architecture which enables the platform to articulate each topic in one thematic module and, above all, to be expanded overtime by adding new modules. In particular, to date, the platform consists of the following 8 modules:

1. Knowledge bases on waste management. It consists of a list of online portals that offer a wide range of information on waste management in different sectors. Stakeholders can search for contents using keywords and filters. The module includes a description of each listed portal (e.g. organization that manages it, sector of interest, purposes and objectives, etc.) and it also includes the different links that create a direct connection to each portal.
2. Guidelines on waste management. The module contains a set of guidelines, accessible free of charge, which represents a valuable support tool for stakeholders in the field of waste management.
3. Research tool through focal areas websites. This tool allows simultaneous searches (by keywords) in all the websites of the leading organizations of the focal areas of the GPWM (e.g. UNEP IETC,[9] ISWA,[10] UNIDO,[11] etc.).
4. Calendar of events on waste management (e.g. seminars, workshops, conferences, meetings, etc.).
5. Descriptive map of recent and ongoing activities on waste management, on a global scale, undertaken by international and intergovernmental organizations.
6. Directory of experts in various fields of waste management related to different organizations, including the leading organizations of the focal areas of GPWM.
7. Country Profiles. The module includes a list of profiles that aim to provide information on waste management in different countries or territories.
8. Evaluation of countries. The module includes a list of analyses aimed at assessing the ability of countries to properly manage waste and scraps. The analyses are based on data and information provided by governments and other institutional stakeholders.

[9]UNEP IETC, International Environmental Technology Centre (Division of UNEP). Website: www.unenvironment.org/ietc.

[10]ISWA, International Solid Waste Association. Website: www.iswa.org.

[11]UNIDO, United Nations Industrial Development Organization. Website: www.unido.org.

8.3 European Union (EU) Contribution: The European Cluster Collaboration Platform (ECCP)

The European Commission is launching a series of initiatives within COSME[12] and Horizon 2020[13]programs to support the innovation and the growth of SMEs through the consolidation of clusters. Furthermore, COM 14/2014 "For a European Industrial Renaissance"[14] highlighted how the creation of SME clusters represents one of the main tools able to facilitate cross-sectoral and cross-border collaboration, supporting in this way the growth and the internationalization of SMEs. As defined by the European Commission, clusters are groups of SMEs and supporting actors that cooperate and collaborate in a particular location.

According to the European Commission,[15] within the clusters SMEs—by collaborating and working together—SMEs can enter into a virtuous network—meant as a promoter and supportive higher entity—which may sustain each of them in:

---

[12]COSME, Competitiveness of Enterprises and Small and Medium-Sized Enterprises, is the European Union program for SMEs (duration: 2014–2020). COSME aims to promote entrepreneurship and to improve the business environment in which SMEs work in order to enable them to fully realize their potential within the global economy. In particular, COSME will support SMEs in: (a) facilitating access to finance; (b) supporting internationalization and market access; (c) creating an environment favorable to competitiveness; (d) encouraging entrepreneurial culture. Source: COSME website: ec.europa.eu/growth/smes/cosme_it.

[13]Horizon 2020 is the framework programme of the European Commission for Research and Innovation that covers the period from 1 January 2014 to 31 December 2020. The three main pillars of the Horizon 2020 programme are:

(a) Scientific excellence. The objective is to increase the quality of the European scientific base, supporting innovative ideas, providing researchers with access to the best research infrastructures and making Europe an attractive place for researchers in the world.
(b) Industrial leadership. The objective is to make Europe a more attractive place to invest in research and innovation by bringing large investments in industrial technologies, fostering the growth of European innovative SMEs—also by giving them the access to adequate levels of financing—and helping them to become world-leading companies.
(c) Challenges of society. The objective is to address the great social concerns bringing together resources from a variety of sectors, technologies and disciplines—including social sciences and humanities—thus constituting a solid knowledge base for political decisions at the European, national and regional levels.

Source: Horizon 2020 website: ec.europa.eu/programmes/horizon2020.

[14]COM 14/2014 "Communication from the Commission to the European Parliament, the Council, the European Economic and Social Committee and the Committee of the Regions. *For a European Industrial Renaissance*" invites EU countries to recognize the importance of a solid industrial base as a key factor for the EU economic growth and the creation of new jobs. COM 14/2014 defines the priorities of the European Commission in the field of industrial policies and it provides an overview of the on-going undertaken actions, aiming at promoting competitiveness, innovation and internationalization and at consolidating and improving the regulatory framework of the industrial field.

[15]European Commission, *European Cluster Collaboration Platform* website: www.clustercollaboration.eu.

- optimizing their practices
- reaching a higher level of innovation;
- revitalizing the economy of their sectors;
- creating more jobs;
- registering more brands and international patents and licenses with respect to what they would be able to do individually.

Among the clusters defined by the European Union, there is the *Waste Management and Recycling Cluster.*[16] The core of the Waste Management and Recycling Cluster (Table 8.3) consists of companies engaged in the collection, treatment, recycling, disposal and transport of any type of industrial waste. In addition, the cluster involves also companies, consulting companies, research institutes and foundations dedicated to environmental education and sensitization.

The main objective of the cluster is to increase the competitiveness of its members, mainly businesses, through:

- the creation of collaboration networks within the waste management sector;
- the promotion and strengthening of the conscious and efficient use of resources and the exchange of knowledge between cluster partners;
- the possibility for cluster companies to introduce innovative products on the market.

To reach these objectives, the ECCP provides cluster members with useful information support tools in order to improve their practices and performance as well as to increase their competitiveness by means of the transnational and international cooperation. In particular, the main technological tool supporting the Waste Management and Recycling Cluster in achieving the objectives described above is the *European Cluster Collaboration Platform* (ECCP).

The ECCP (Fig. 8.2) is a tool for sharing knowledge between clusters that allows the engaged members to:

- efficiently use network tools in order to search and find potential partners and new business opportunities;
- develop European and international win-win collaborations;
- support the emergence of new value-chains through inter-sectorial cooperation;
- access to reliable and updated information on cluster development;
- improve the cluster performance and the competitiveness of its members.

By providing a valuable set of services and tools as well as creating a virtual place for collaboration and sharing, the ECCP aims to become the main European hub for international cluster cooperation. Indeed, the ECCP has the potential to create cluster bridges between Europe and the world, being able to generate a supportive and encouraging environment for the cooperation of the involved stakeholders.

In particular, the ECCP platform is able to provide matchmaking services for clusters, such as meetings, congresses, conferences, seminars, etc., organized both

[16]*Waste Management and Recycling Cluster* website: www.clustercollaboration.eu/cluster-organisations/waste-management-and-recycling-cluster.

Table 8.3 General data and information about the Waste Management and Recycling Cluster. Adapted from: Waste Management and Recycling Cluster website

Waste management and recycling cluster	
Date of establishment	29 May, 2011
Legal form of the cluster organization	Non-profit
EU priority areas	– Waste management – Waste collection, treatment and disposal activities – Materials recovery and remediation activities
Total number of members	90
Number of SME members	52
Number of universities/research centers	30
Mission/objectives	The Waste Management and Recycling Cluster aims to increase competitiveness of the businesses, institutions, organizations operating in the waste management industry and related industries. To reach this goal the key is the sharing of possessed resources and the exchange of knowledge. The Cluster offers the possibility of concretely implementing innovative products on the market
Support services provided	– Facilitation of collaboration between members – Support of knowledge transfer – Support of technology transfer – Innovation management/support of innovation processes (internal, external) – Development of the region's knowledge-based economy – Legal, marketing and technology consulting – Consulting regarding financial aid for the promotion of activities, including funds for developing collective projects from UE and other funds – Training, seminars and sector conferences as well as participation in both national and international meetings and discussion panels

in Europe and internationally. Moreover, the ECCP offers a set of digital (or virtual) services including:

- dynamic mapping of organizations of the cluster worldwide, therefore, updated mapping of the companies profiled in the ECCP. This dynamic mapping is associated with the possibility of searching for organizations, based on advanced filters (Fig. 8.3).
- provision of detailed information and reports about the European Strategic Cluster Partnerships (ESCP).[17]

[17]The European Strategic Cluster Partnerships (ESCP) have been launched by the European Commission under COSME programme with the aim of encouraging clusters from Europe to

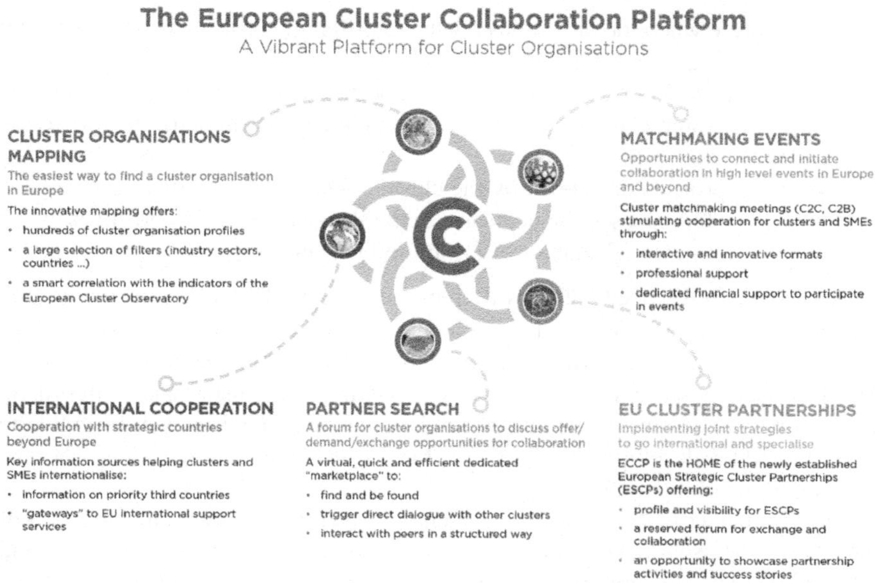

Fig. 8.2 Main features of the European Cluster Collaboration Platform (ECCP). *Source* ECCP 2018 (Presentation at TAFTIE annual event vilnius 26th September 2018, available at: www.clustercollaboration.eu)

Fig. 8.3 ECCP Cluster Organizations Mapping Tool. *Source* https://mapping.clustercollaboration.eu

- a marketplace where organizations related to different clusters can communicate, share and exchange their demands and offers.
- list and description of associations and support organizations useful for clusters such as, for example, organizations that sustain the internationalization processes of SMEs.
- database of projects developed by the clusters within the various European programmes, including a section dedicated to the showcasing of best practices which may represent potential sources of inspiration for new projects.

According to the needs of the different clusters, the ECCP shapes the offered services overtime, establishing itself as a high-performance information platform—dynamic and up-to-date—always attentive to the needs of its members. Indeed, the ECCP is at the service of the organizations belonging to the clusters and it offers them a wide range of services and tools useful for creating an environment favorable to the birth and the development of win-win long-term collaborations among clusters, creating bridges among them.

Moreover, the ECCP information contents are useful, not only for the members of the clusters themselves, but also for the decision makers at the regional, national and international level. Therefore, the platform is connected and open to all interested parties, on a global scale.

8.4 Horizon 2020 Projects of Smart Waste Management

In this section some virtuous projects of Smart Waste Management, funded by the European Commission under the Horizon 2020 Programme,[18] are introduced and discussed. In particular, the investigation was carried out as a thematic research of EU funded projects on the public portal of the European Commission CORDIS (Community Research and Development Information Service).[19] The investigation has highlighted the strong interest of the EU to the projects aimed at optimizing and innovating—by means of the ICTs application—the current waste management

enhance cross-regional and cross-sectoral collaboration. The EU cluster partnerships are expected to gather and share resources and knowledge for collaborating and co-working on joint strategies of common interest for their SME members. Source: https://www.clustercollaboration.eu/eu-cluster-partnerships.

[18]Horizon 2020 is the framework program of the European Commission for research and innovation. It covers the period from 1 January 2014 until 31 December 2020. Website: ec.europa.eu/programmes/horizon2020.

[19]CORDIS, the Community Research and Development Information Service, is the main archive and public portal of the European Commission aimed at disseminating information concerning all EU-funded research projects and their results. The website includes all public information held by the European Commission (e.g. project information sheets, publishable reports and results, editorial content, etc.) to support the communication and enhancement of projects as well as the connections and links to external complementary information sources (e.g. websites, open access publications, etc.). Website: cordis.europa.eu.

practices. In particular, EU is supporting virtuous initiatives and projects dealing with the innovation of waste management practices, covering the following research and action areas:

- addressing, facilitating and promoting the shift towards a circular economy through structured processes of industrial symbiosis, articulated according to different scales of application (e.g. district, city, regional, national, European, etc.);
- developing a shared systematic approach (e.g. common strategies, methodologies, procedures, tools, etc.) for the reduction, recycling and reuse of waste (with a focus on the recycling of raw materials from products and buildings) at European and global scale;
- developing strategies aimed at moving towards near-zero waste at European and global scale.

The number of funded project related to the waste priority is continuously increasing. Among the multiplicity of waste management-related projects, some virtuous examples are described and analyzed below (Table 8.4).

Table 8.4 Examples of Smart Waste Management projects funded by EU under Horizon 2020 Programme

	Project	Coordinating country	Start—end dates	Overall budget	EU contribution
1	**Urban_Wins.** Urban metabolism accounts for building waste management Innovative Networks and Strategies	Italy	From 01/06/2016 to 31/05/2019	€ 496 6516,25	€ 496 6516,25
2	**Waste4Think.** Moving towards life cycle thinking by integrating advanced waste Management Systems	Spain	From 01/06/2016 to 30/11/2019	€ 10 521 412,35	€ 8 818 556,12
3	**BAMB.** Buildings as material banks: integrating materials passports with reversible building design to optimize circular industrial value chains	Belgium	From 01/09/2015 to 31/08/2018	€ 9 950 388,75	€ 8 858 763,02

(continued)

Table 8.4 (continued)

	Project	Coordinating country	Start—end dates	Overall budget	EU contribution
4	**FISSAC**. Fostering industrial symbiosis for a sustainable resource intensive industry across the extended construction value chain	Spain	From 01/09/2015 to 29/02/2020 On-going project	€ 11 523 404,81	€ 9 108 594,25
5	**RUBSEE**. Extending artificial intelligence revolution in the waste field beyond sorting	Spain	From 01/02/2017 to 30/09/2019	€ 1 800 176,75	€ 1 260 123,73
6	**SmartWASTE**. Smart logistics for WASTE and recycling operations in European cities	Finland	From 01/06/2016 to 31/05/2019	€ 2 100 449,56	€ 1 470 314,70

1. **Urban_Wins**. Urban metabolism accounts for building Waste Management Innovative Networks and Strategies

References

https://cordis.europa.eu/project/rcn/203270/factsheet/en
https://www.urbanwins.eu/

Coordinated in

Italy

EU Programme

H2020-EU.3.5.4.—Enabling the transition towards a green economy and society through eco-innovation

EU Topic

WASTE-6b-2015—Eco-innovative strategies

Start—end dates

From 01/06/2016 to 31/05/2019

Overall budget

€ 496 6516,25

EU contribution

€ 496 6516,25

Funding Scheme

RIA—Research and Innovation action

Coordinator

Comune di Cremona (Public body)

Project description

"*The aim of the project is to develop and test methods for designing and implementing innovative and sustainable Strategic Plans for Waste Prevention and Management in various urban contexts that will enhance urban environmental resilience and guarantee progress towards more sustainable production and consumption patterns together with improvements waste recovery and recovered materials use. Urban_Wins will define a data set, based on material flow indicators, capable of supporting and orienting decision making processes for urban waste prevention and management. Knowledge of the factors that influence the metabolism of cities will be improved together with the understanding of how those factors can be transformed in positive drivers of technological, non-technological and governance changes. The information set produced by the consortium will also focus on how a more efficient use of resources and a better management of waste can improve urban quality and citizens' welfare, key points for urban stakeholder's involvement, both in the planning and implementation of actions. The proposal reunites diverse actors such as cities, research institutes and universities, environmental NGOs, IT&C, technological innovation and waste management companies, professional associations that represent EU regions, sectors and levels of governance. The complex partnership guarantees that advancement in EU research in the field of urban metabolism and waste management strategies is directly linked to stakeholder engagement and mutual learning and contributes to the achievement of resource efficiency and waste management objectives. Urban_Wins analytical tools will be built on the base of datasets and experiences of 24 EU cities from 6 European countries and the Strategic Plans will be tested by 8 EU cities and will encompass regulatory measures, educational initiatives and sector specific actions*". Source: https://cordis.europa.eu/project/rcn/203270/factsheet/en

2. **Waste4Think**. Moving towards Life Cycle Thinking by integrating Advanced Waste Management Systems

References

https://cordis.europa.eu/project/rcn/203385/factsheet/en
http://waste4think.eu/

Coordinated in

Spain

EU Programme

H2020-EU.3.5.4.—Enabling the transition towards a green economy and society through eco-innovation

EU Topic

WASTE-6a-2015—Eco-innovative solutions

Start—end dates

From 01/06/2016 to 30/11/2019

Overall budget

€ 10 521 412,35

EU contribution

€ 8 818 556,12

Funding Scheme

IA—Innovation action

Coordinator

Fundacion Deusto (Research organization)

Project description

"The main objective of this project is to move forward the current waste management practices into a circular economy motto, demonstrating the value of integrating and validating a set of 20 eco-innovative solutions that cover all the waste value chain. The benefits of these solutions will be enhanced by a holistic waste data management methodology, and will be demonstrated in 4 complementary urban areas in Europe.
The eco-innovative solutions include technological and non-technological tools such as:

(a) *IT tools to support the daily operation and long-term planning;*
(b) *Apps for citizens empowerment and engagement;*
(c) *Educational materials based on innovative teaching units and serious games;*
(d) *Tools for citizen science for the co-creation of novel solutions;*
(e) *Mechanisms to boost behavioral changes based on economic instruments and social actions;*
(f) *Decentralized solutions for valorization and reuse of high value resources.*

The different solutions will be implemented in 4 complementary European areas:

(a) *Zamudio (ES) is a highly industrialized area with a spread population that uses a separated kerbside collection;*
(b) *Halandri (GR) is a large suburban city with a wide range of business that has a very basic waste management system;*
(c) *Seveso (IT) is a residential town that uses a door-to-door system;*
(d) *and Cascais (PT) is an extensive and high touristic coastal town that implements an advanced collection system.*

The project includes a consortium of 19 partners with 4 public agencies and administrations, 3 research centers and universities, 8 SMEs, 2 LEs, 1 cluster and 1 NGO". Source: https://cordis.europa.eu/project/rcn/203385/factsheet/en

3. **BAMB**. Buildings As Material Banks: integrating materials passports with reversible building design to optimize circular industrial value chains

References

https://cordis.europa.eu/project/rcn/196829/factsheet/en
https://www.bamb2020.eu/

Coordinated in

Belgium

EU Programme

H2020-EU.3.5.4.—Enabling the transition towards a green economy and society through eco-innovation

EU Topic

WASTE-1-2014—Moving towards a circular economy through industrial symbiosis

Start—end dates

From 01/09/2015 to 31/08/2018

Overall budget

€ 9 950 388,75

EU contribution

€ 8 858 763,02

Funding Scheme

IA—Innovation action

Coordinator

Institut Bruxellois pour la Gestion de l'Environnement—Brussels Instituut Voor Milieubeheer (Education establishment)

Project description

"The aims of BAMB (Buildings as Material Banks) are the prevention of construction and demolition waste, the reduction of virgin resource consumption and the development towards a circular economy through industrial symbiosis, addressing the challenges mentioned in the Work Programme on Climate action, environment, resource efficiency and raw materials. The focus of the project is on building construction and process industries (from architects to raw material suppliers).

The BAMB-project implements the principles of the waste hierarchy: the prevention of waste, its reuse and recycling. Key is to improve the value of materials used in buildings for recovery. This is achieved by developing and integrating two complementary value adding frameworks, (1) materials passports and (2) reversible building design.

These frameworks will be able to change conventional (cradle-to-grave) building design, so that buildings can be transformed to new functions (extending their life span) or disassembled to building components or material feedstock that can be up-cycled in new constructions (using materials passports). This way, continuous loops of materials are created while large amounts of waste will be prevented.

Activities from research to market introduction are planned. Fundamental knowledge gaps should be bridged in order to introduce both frameworks on the market. Advanced ICT tools and management models will enable market uptake and the organization of circular value chains in building and

process industries. New business models for (circular) value chains will be developed and tested on selected materials.
The inclusion of strategic partners along the value chains in an industrial board will maximize market replicability potential, while several (mostly privately funded) building pilots will demonstrate the potential of the new techniques. Awareness will be raised to facilitate the transition towards circularity by policy reform and changing consumer behavior". Source: https://cordis.europa.eu/project/rcn/196829/factsheet/en

4. **FISSAC**. Fostering industrial symbiosis for a sustainable resource intensive industry across the extended construction value chain

References

https://cordis.europa.eu/project/rcn/196821/factsheet/en
http://fissacproject.eu/it/

Coordinated in

Spain

EU Programme

H2020-EU.3.5.4.—Enabling the transition towards a green economy and society through eco-innovation

EU Topic

WASTE-1-2014—Moving towards a circular economy through industrial symbiosis

Start—end dates

From 01/09/2015 to 29/02/2020 On-going

Overall budget

€ 11 523 404,81

EU contribution

€ 9 108 594,25

Funding Scheme

IA—Innovation action

Coordinator

ACCIONA CONSTRUCCION SA

Project description

"The overall objective of FISSAC project is to develop and demonstrate a new paradigm built on an innovative industrial symbiosis model towards a zero waste approach in the resource intensive industries of the construction value chain, tackling harmonized technological and non technological requirements, leading to material closed-loop processes and moving to a circular economy.

A methodology and a software platform will be developed in order to implement the innovative industrial symbiosis model in a feasible scenario of industrial symbiosis synergies between industries (steel, aluminium, natural stone, chemical and demolition and construction sectors) and stakeholders in the extended construction value chain. It will guide how to overcome technical barriers and non technical barriers, as well as standardization concerns to implement and replicate industrial symbiosis in a local/regional dimension. The ambition of the model will be to be replicated in other regions and other value chains symbiosis scenarios. The model will be applied based on the three sustainability pillars.

FISSAC will demonstrate the applicability of the model as well as the effectiveness of the innovative processes, services and products at different levels:

- *Manufacturing processes: with demonstration of closed loop recycling processes to transform waste into valuable secondary raw materials, and manufacturing processes of the novel products at industrial scale.*
- *Product validation: with demonstration of the eco-design of eco-innovative construction products (new Eco-Cement and Green Concrete, innovative ceramic tiles and Rubber Wood Plastic Composites) in pre-industrial processes under a life cycle approach, and demonstration at real scale in different case studies of the application and the technical performance of the products.*
- *FISSAC model, with the demonstration of the software platform and replicability assessment of the model through living lab concept".* Source: https://cordis.europa.eu/project/rcn/196821/factsheet/en

5. **RUBSEE**. Extending artificial intelligence revolution in the waste field beyond sorting

References

https://cordis.europa.eu/project/rcn/209094/factsheet/en
http://sadako.es/rubsee-project/

Coordinated in

Spain

EU Programmes

H2020-EU.3.5.—SOCIETAL CHALLENGES—Climate action, Environment, Resource Efficiency and Raw Materials
H2020-EU.2.3.1.—Mainstreaming SME support, especially through a dedicated instrument

EU Topic

SMEInst-11-2016-2017—Boosting the potential of small businesses in the areas of climate action, environment, resource efficiency and raw materials

Start—end dates

From 01/02/2017 to 30/09/2019

Overall budget

€ 1 800 176,75

EU contribution

€ 1 260 123,73

Funding Scheme

SME-2—SME instrument phase 2

Coordinator

SADAKO TECHNOLOGIES SL

Project description

"Current WTPs (Waste Treatment Plants) aren't able to recover all the valuable waste they process, indeed more valuable materials are lost and landfilled or incinerated. The reason of this wasteful spending is clear: current methods do not allow an increase in material recuperation in a cost-effective way: the incremental cost of recovering more materials is bigger than the market value of the additional materials recovered.

Current technologies aren't enough to meet EU regulations like directive 2008/98/EC, which requires that 50% of household waste is recovered by 2020.
SADAKO has developed RUBSEE, a disruptive real-time monitoring system (using Computer vision + Artificial intelligence) of waste flows in a WTP in order to optimize the performance/operation thereof and the recovery of different materials.
RUBSEE will allow waste industry improve its economic, regulatory compliance and environmental performance with a solution that is cost efficient and complementary to actual solutions.
Thanks to RUBSEE data, current equipment performance can be improved up to 20% by adapting their parameters to the variability of the waste flow on real time". Source: https://cordis.europa.eu/project/rcn/209094/factsheet/en

6. **SmartWASTE**. Smart logistics for WASTE and recycling operations in European cities

References

https://cordis.europa.eu/project/rcn/204352/factsheet/en
https://www.enevo.com/smartwaste/

Coordinated in

Finland

EU Programmes

H2020-EU.3.4.—SOCIETAL CHALLENGES—Smart, Green And Integrated Transport
H2020-EU.2.1.1.—INDUSTRIAL LEADERSHIP—Leadership in enabling and industrial technologies—Information and Communication Technologies (ICT)
H2020-EU.2.3.1.—Mainstreaming SME support, especially through a dedicated instrument EU Topic
SMEInst-11-2016-2017—Boosting the potential of small businesses in the areas of climate action, environment, resource efficiency and raw materials

Start—end dates

From 01/06/2016 to 31/05/2019

Overall budget

€ 2 100 449,56

EU contribution

€ 1 470 314,70

Funding Scheme

SME-2—SME instrument phase 2

Coordinator

ENEVO OY (Private for-profit entity)

Project description

"The key problem in waste collection today is static routes and schedules: truck drivers are driving "blindly" from bin to bin and collecting containers that are either half empty or over filled. This adds up to a large amount of unnecessary costs, such as time spent, gas consumption and greenhouse gas emissions. Globally, over 400 million waste containers are being served by millions of trucks every day, and 50 % of the value in the market is in the logistics.
Enevo is a growing Finnish technology company that aims to capitalize on this 12-billion-euro business opportunity and become the #1 supply chain platform company for waste and recycling operations worldwide. As waste management plays a central role in the circular economy, Enevo is developing more efficient waste collection and management systems. Enevo helps its customers make their waste and recycling operations more efficient, leading to a more sustainable world.
The vision of Enevo is to turn all waste in the world into a valued resource. SmartWASTE project is addressing two significant EU-wide challenges:

(1) *Optimizing transport operations and tackling the environmental and logistical challenges that the European transport sector is facing*
(2) *Waste management in the circular economy context.*

The objective of SmartWASTE proposal is to scale-up and expand the service into new European regions by piloting the solution with potential customers in 10 large scale pilots. Through piloting, Enevo gains important feedback that is provided back to product development to improve Enevo offering and operations to be better suited for large scale regional expansion". Source: https://cordis.europa.eu/project/rcn/204352/factsheet/en

The conducted analysis shows applications of ICT solutions and digital information platforms to improve processes, methods, tools and procedures—both at a strategic and operational level—of scraps/waste management at different scales (e.g. single building, district, city, production system, etc.). This attention is also demonstrated by the favorable amount of funds given by the EU for the implementation of these projects.

In particular, it is possible to identify the following areas of interest shared by the analyzed projects:

- development and promotion of cost-efficient and energy-efficient eco-innovative technologies, processes and/or services for waste prevention and for the recycling and recovery of high-grade materials from waste;
- development and promotion of "waste-resource" processes of industrial symbiosis (stakeholders derive mutual benefit from sharing scraps/waste materials since the scraps/waste of one industry can be a resource for another one). However, the implementation of these virtuous processes needs proper information management tools (e.g. information platform) for the collection and processing of waste-related data (e.g. in order to understand the quantity of waste owned by companies, to link different companies with possible mutual interests, to perform estimations of possible industrial symbiosis scenarios, etc.).

Concluding, the topic of ICT-based Waste Management or Smart Waste Management proves nowadays to be a hot topic with many scenarios of improvement, optimization and innovation that can be successfully carried out if properly supported by: (a) funding bodies, (b) shared information references and tools, (c) the extensive coordination among a variety of stakeholders (e.g. industry, civil society organizations, research, policy makers, public authorities, etc.)—engaged into a collaborative network, sharing knowledge and experience—as well as, above all, (d) the greater awareness among companies of the responsibilities linked to waste generation, which is a crucial aspect given the central role of companies with respect to the current and future economic and social transformation.

References

1. UNEP (2011) Towards a green economy. Pathways to sustainable and poverty eradication. A synthesis for policy makers. Available at: https://sustainabledevelopment.un.org/content/documents/126GER_synthesis_en.pdf. Accessed in April 2019
2. UNEP 2011GPWM. Framework Document. Available at: http://www.unep.or.jp/Ietc/SPC/activities/GPWM/GPWMFrameworkDocumentv.11282011.pdf. Accessed in April 2019
3. Talamo C, Migliore M (2017) Le utilità dell'inutile. Economia circolare e strategie di riciclo dei rifiuti pre-consumo per il settore edilizio. Maggioli
4. Costa I, Massard G, Agarwal A (2010) Waste management policies for industrial symbiosis development: case studies in European countries. J Clean Prod 18(8):815–822
5. Ntasiou M, Andreou E (2017) The standard of industrial symbiosis: environmental criteria and methodology on the establishment and operation of industrial and business parks. Proced Environ Sci 38:744–751
6. ISO/IEC JTC 1 (2015) Big data. Preliminary Report 2014. Available at https://www.iso.org/files/live/sites/isoorg/files/developing_standards/docs/en/big_data_report-jtc1.pdf. Accessed in April 2019
7. Sun Z (2018) 10 bigs: big data and its ten big characteristics. PNG UoT BAIS 3(1):1–10
8. Vorhies W (2014) How many "V's" in big data? The characteristics that define big data. Data Science Central. Available at: https://www.datasciencecentral.com/profiles/blogs/how-many-v-s-in-big-data-the-characteristics-that-define-big-data. Accessed in April 2019

9. Talamo C, Atta N, Martani C, Paganin G (2016) L'integrazione delle infrastrutture urbane fisiche e digitali: il ruolo dei "Big Data". TECHNE—J Technol Archit Environ 11:217–225
10. Wang L (2017) Heterogeneous data and big data analytics. Autom Contr Inf Sci 3(1):8–15
11. Bina S, Zhiquana Y, Jonathana LSC, Jieweia DK, Kurleb D, Cerdasb F, Herrmannb C (2015) A big data analytics approach to develop industrial symbioses in large cities. In: Proceedings of the 22nd CIRP conference on life cycle engineering, CIRP 2015, pp 450–455
12. Ellen MacArthur Foundation, McKinsey Centre for Business and Environment, Stiftungsfonds für Umweltökonomie und Nachhaltigkeit—SUN (2015) Growth within: a circular economy vision for a competitive Europe. Available at: https://www.ellenmacarthurfoundation.org/assets/downloads/publications/EllenMacArthurFoundation_Growth-Within_July15.pdf. Accessed in April 2019
13. EU (2010) COM 2020/2010. Communication from the commission Europe 2020 A strategy for smart, sustainable and inclusive growth
14. EU (2014) COM 14/2014. Communication from the commission to the European Parliament, the council, the European economic and social committee and the committee of the regions. For a European Industrial Renaissance
15. EU (2014) COM 398/2014. Communication from the commission to the European Parliament, the council, the European economic and social committee and the committee of the regions. Towards a circular economy: a zero waste programme for Europe
16. EU (2015) COM 614/2015. Communication from the commission to the European Parliament, the council, the European economic and social committee and the committee of the regions. Closing the loop—an EU action plan for the Circular Economy
17. Global Partnership on Waste Management—GPWM (2016) Framework Document, 22 June 2016. Available at: http://www.sids2014.org/content/documents/commitments/9501_7462_commitment_The%20GPWM%20Framework%20Document.pdf. Accessed in April 2019
18. IBM Sales and Distribution December (2015) White Paper. IBM intelligent waste management platform. Using the power of analytics to profit from the move to a circular economy. Available at: https://www.ibm.com/downloads/cas/41Q1DYO6. Accessed in April 2019
19. Iraldo P, Bruschi I (2015) Economia circolare: principi guida e casi studio. Osservatorio sulla Green Economy, IEFE Bocconi
20. Shahrokni H, Van der Heijde B, Lazarevic D, Brandt N (2014) Big data GIS analytics towards efficient waste management in stockholm. In: Proceedings of the 2nd International Conference on ICT for Sustainability ICT4S 2014, pp 140–147

Websites

21. https://cordis.europa.eu
22. https://cordis.europa.eu
23. https://cordis.europa.eu/project/rcn/196821/factsheet/en
24. https://cordis.europa.eu/project/rcn/196829/factsheet/en
25. https://cordis.europa.eu/project/rcn/203270/factsheet/en
26. https://cordis.europa.eu/project/rcn/203385/factsheet/en
27. https://cordis.europa.eu/project/rcn/204352/factsheet/en
28. https://cordis.europa.eu/project/rcn/209094/factsheet/en
29. https://ec.europa.eu/programmes/horizon2020
30. www.clustercollaboration.eu
31. www.clustercollaboration.eu
32. www.datasciencecentral.com
33. www.ec.europa.eu
34. www.ellenmacarthurfoundation.org
35. www.gartner.com/it-glossary

36. www.ispra.it
37. www.iswa.org
38. www.itu.int
39. www.unenvironment.org

Chapter 9
Integrated Design and Living Labs to Fostering Smart (Waste) Networks

Abstract There is increasing concern about environmental sustainability and waste management issues. This study argues that the combination of process integration and participatory organization can support and promote the creation of smart networks aimed at improving the sustainability performance at both the building and urban levels. This work is intended to act as a starting point to theoretically connect the two bodies of literature: on one side, integrated design process and integrated project delivery, on the other side, participatory organization and the Living Lab approach. Little studies have been done to couple the two topics as a strategy to foster and facilitate smart waste networks in the context of the building sector. To achieve this, first, the Chapter reviews relevant literature about the integrated design process and integrated project delivery. The Chapter proceeds then with presenting literature about participatory organization and the Living Lab approach. Lastly, the Chapter concludes with the presentation of a proof of concept of collaborative service aimed at boosting social innovation and smart waste networks and developed by applying an integrated design process and a Living Lab approach. Some final considerations are proposed.

Keywords Integration · Integrated design · Participation · Living lab · Smart waste network · Sustainability

9.1 From Relationships to Partnerships: A Paradigm Shift to-wards Integrated Building Design and Delivery

There is increasing concern about the lack of efficiency of the building processes as well as the low quality of the building products from the perspective of sustainability and waste management issues (e.g. Fulford and Standing [1]). This can be attributed to the lack of integration between project stakeholders since the early stages of

This chapter is authored by Marcella Bonanomi, Chair of Innovative and Industrial Construction, Institute of Construction and Infrastructure Management, ETH Zurich. e-mail: bonanomi@ibi.baug.ethz.ch.

M. Migliore et al., *Strategies for Circular Economy and Cross-sectoral Exchanges for Sustainable Building Products*, Springer Tracts in Civil Engineering,
https://doi.org/10.1007/978-3-030-30318-1_9

the building process, which also translates into a low information and knowledge exchange throughout the entire process [2].

Today, many calls exist aimed at improving processes' efficiency and products' quality. The rationale of most of them is boosting innovation through more *integration* in both the design process and project delivery [3, 4]. However, the AEC sector is well-known for low levels of innovation adoption (e.g., Noktehdan [5]). This is due to the one-of-a-kind nature of building projects that are typically organized by (dis)integrating the work between many stakeholders [6]. Furthermore, traditional delivery methods, such as the commonly used Design-Bid-Build (DBB) method, which are still adopted broadly in Italy and beyond, provide a contractual basis that is conducive to silo-based and sequential processes, and to fragmented supply chains [7]. Within these conventional contracting configurations, projects are usually managed as temporary multi-organizations settings in which each party acts as an autonomous unit focusing on its own interests and incentives [8]. In most cases, conflict and adversarial behaviors characterize relationships among the three main project parties: owners/clients, contractors, and designers [9]. Moreover, after project delivery, both the designers and contractors are no longer responsible for the long-term building performance. The responsibility is placed at the owner side, yet there are no incentives for the other parties to care about the building life-cycle in terms of sustainability performance.

This lack of integration results in a building sector that still struggles to make productivity and quality gains in comparison to other industries, such as manufacturing [3]. On the other side, the development and adoption of digital technologies and processes, such as Building Information Modeling (BIM), creates a demand for more integrated processes, information, and organizations [10–12]. Accordingly, researchers argue for the benefits of more integrated and relational scenarios resembling the way ecosystems and networks work [13].

As far back as 1988, Duhon [14] defined knowledge management as "A discipline that promotes an integrated approach to identifying, capturing, evaluating, retrieving, and sharing all of an enterprise's information assets". In more recent years, Hegazy and Ghorab [15] argue that "The new paradigm is that knowledge must be shared for it to grow". Based on this, integration processes and practices should be promoted not only within firms, but also across business boundaries through the development of inter-organizational and cross-functional networks and communities of practice aimed at fostering an *integrated design and delivery*.

A recent report by Accenture [16], which proposes the concept of a digital supply network, states: "In a metaphorical sense, people and data -as well as materials, products and supplies—must travel together across the extended enterprise". This shift from supply chains to supply networks does not therefore affect only business processes, but also people behaviors and data streams.

Based on this network paradigm, new innovations in project delivery are emerging. Today, in fact, some projects are delivered using Multi-Party Construction Arrangements (MPCAs), such as Integrated Project Delivery (IPD) [9, 17]. MPCAs provide a new legal and management framework that supports project stakeholders' integration

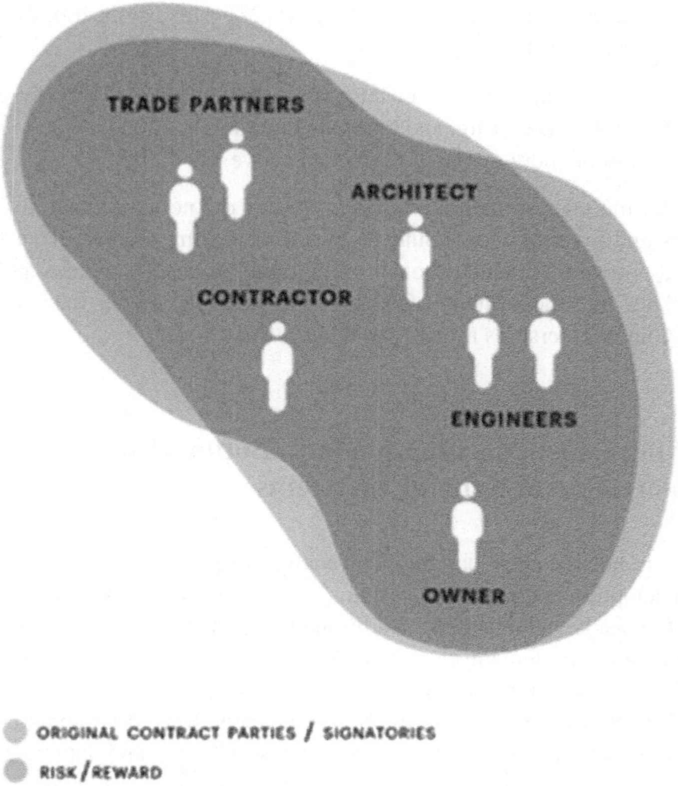

Fig. 9.1 Example of project stakeholders' integration in the context of an IPD project [19]

and information and knowledge exchange since the early phases of the building process [18]. MPCAs seek therefore to prioritize collaborative partnerships rather than adversarial relationships [19] (see Fig. 9.1). Additionally, recent quantitative and comparative case study research demonstrates that MPCAs significantly improve productivity and result in better cost, schedule, and performance of design and construction projects [20–23].

Given these considerations, it should be underlined that all the project parties should facilitate the creation of smart networks in order to move towards an integrated design and delivery. Technology providers, for example, could achieve this integration and "networkification" goal by:

1. Developing integrated technology ecosystems,
2. Strengthening the interoperability of their products,
3. Striving for open standards to address critical interfaces between software and tools.

Other stakeholders, such as owners/clients, must make appropriate adjustments as well. Today, in fact, owners/clients are increasingly asked to play a key role in the design and management of the building process [24]. This can be translated into the following double way strategy:

- Design firms should promote an integrated design and delivery by enhancing knowledge creation, sharing, and communication with clients/owners;
- Clients should be more prepared to play an active role [25].

Lastly, for the sake of integration and in the age of "networkification" [26], both researchers and practitioners should be aware of the importance of an interdisciplinary and cross-functional approach to the building design process and delivery. This translates into the strategy of fostering interactions, rather than transactions, and facilitating partnerships rather than relationships.

9.2 Living Labs: Participation for a Bottom-Up Co-creation of Urban Ecosystems

Integration is not the only paradigm of change to be considered for the sake of innovation adoption and improved sustainability performance in the building sector: *participation* is a key element too. Additionally, the building system is not the only level at which to implement changes: the boundaries need to be enlarged at the urban level. Contemporary cities, in fact, moved away from a static urban scheme, articulated in physical spaces and mobility channels, toward complex ecosystems of physical and digital spaces where mixed flows of people and data interact one to another [27]. However, although the deep transformation that contemporary cities are facing by being increasingly net-worked and articulated into digital flows and immaterial connections, management and planning of urban systems are still blocked by traditional top-down methods and analogic tools.

Living Labs (LL) are an emerging approach that aims at fostering the bottom-up co-creation of innovative and smart urban ecosystems. In recent years, Living Labs have become a powerful approach to effectively integrate the end-user of a service at all stages of the research, development, and innovation process. Peltomaki [28] defines Living Labs as "open innovation environments in real life settings, in which user-driven innovation is fully integrated within the co-creation process of new services, products, and societal infrastructures." The main objective of a Living Lab is to boost an open innovation ecosystem within which users can play a fundamental role in the research, development, and innovation processes. The stakeholders, who are typically involved in open innovation contexts and Living Lab approaches, can be described through a 3Ps networked structure: Public (governmental authorities, academic institutions, and welfare), Private (business partnership and no profit association) and People (local communities and virtual communities).

The LL approach is focused on three fundamental steps:

1. Involving users into the creative process in order to easily individuate new behaviors and patterns;

2. Filling the gap between technology development and the uptake of new services and products involving the public-private-people partnership;
3. Providing validity to new technological solutions by early demonstrating their socio-economic implications.

The Living Lab approach is built on the idea that users are considered "innovators", hence it is important to get access to their ideas and knowledge through an interactive, co-creative, and top-down process. It is an approach that aims to fulfill users' needs by placing people at the center of the innovation ecosystem. The Living Lab is a real-life setting open environment, where this potential innovation is an integral part of the co-designing process of new services and products. Living Labs could fulfill their roles, providing the necessary outputs for the creation of platforms where the interactive and co-creative process of innovation can take place only when the users are considered innovators and their knowledge and ideas are primarily targeted. Combining Living Labs with the user-centered innovation focus predicts a unique market, where a spectrum of experiences can be collected and filtered to build new knowledge in the innovation domain.

9.3 The &CO Project: A Collaborative Service to Boosting Smart Waste Networks

The "&CO Project-Use Everything, Waste Nothing" is the proof of concept of a collaborative service, which aims at connecting two different, but potentially complementary urban communities: the one of waste materials suppliers and the one of raw materials users. More in details, the &CO Project aims at positively effecting the huge waste of materials. In fact, everyday a significant amount of materials, which might still have an economical and use value, are thrown away.

The &CO team did not accept the common idea that sustainability means only recycling, but rather opted for another philosophy: use everything, waste nothing. Therefore, the core concept of the &CO Project is the idea that the discard of someone can be a raw material for someone else. The solution lies in transforming what appears as a problem in a potential value: it is possible to reduce the waste of materials by extending their life cycles and their use-phases through an up-cycling process. In particular, the collaborative service envisioned by the &CO project aims at networking, through a digital platform—a free application for smartphones—all the possible suppliers of waste materials (copy centers, box factories, supermarkets, carpentries, etc.) with the ones willing to reuse them (universities, cultural associations, art schools, etc.) (see Fig. 9.2).

The final goal of the service has been individuated by adopting a Living Lab approach: all the different social actors, students, residents, commuters, who daily live the city,[1] have been involved (see Fig. 9.3). This approach has led first to the

[1]Milano, and the 'Città Study' area and the 'Leonardo' campus of the university Politecnico di Milano in particular, have been selected as fields of action for the development of the proof of concept of the and CO project.

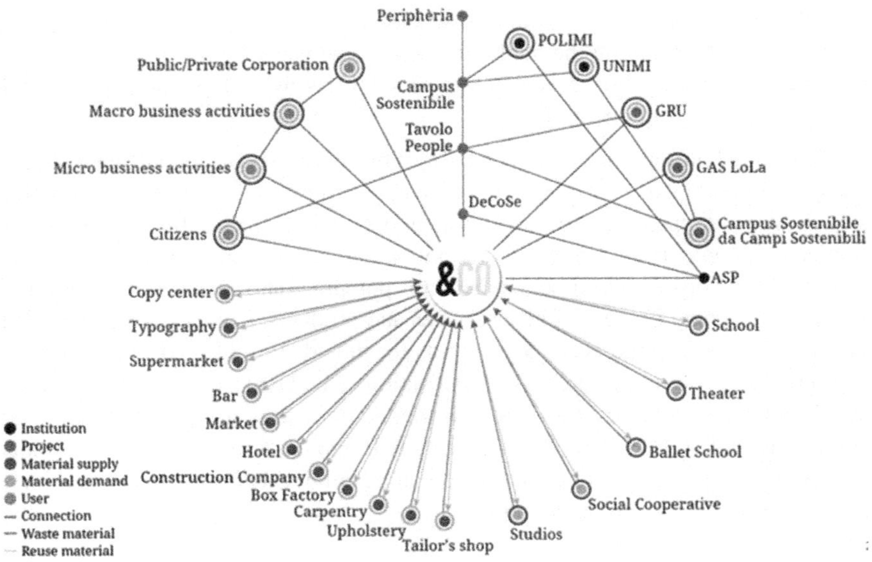

Fig. 9.2 Potential stakeholders' network of the &CO Project [29]

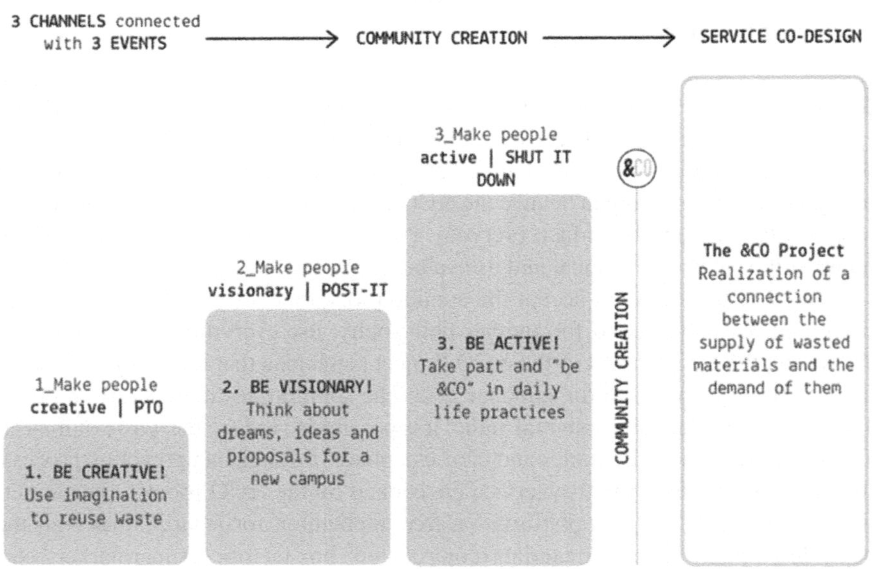

Fig. 9.3 Strategy of people engagement for the &CO project [29]

identification of what was really considered as an unsustainable behavior in the Campus by the users of the area and therefore what could be a shared focus on which to work about. Secondly it has led to the engagement of an active community willing to be part of an integrated design solution process.

9.4 Consideration

There is increasing concern about unsustainable choices and behaviors and the associated environmental impact. Today many calls exist, such as the ones from the European Union, which aim at fostering more sustainable processes and products. In particular, the reduction of waste has been identified as a key strategy. This translates into a demand to initiate a process of transformation in which environmental sustainability, and waste reduction in particular, couples with process and product innovation.

To achieve this, this study proposes two paradigms of change: integration and participation. On one hand, the building design process should become more integrated in order to enable project stakeholders to input their specific knowledge since the early phases of the process. Integration can also be supported by the use of digital technologies and processes, such as Building Information Modeling (BIM). BIM, in fact, can help to take data-informed choices in terms of sustainability and waste reduction goals. Additionally, project delivery and management processes should change accordingly in order to facilitate the creation of cross-functional networks, which can help to achieve better sustainability performance of construction projects.

On the other hand, participation of people should be fostered at the urban level through innovative approaches, such as the Living Lab method, in order to achieve socio-technological innovation. The &CO Project clearly exemplifies a collaborative service in which people and technology have been merged into a complex system to give solutions to some current problems and make meaningful steps towards a more sustainable society.

References

1. Fulford R, Standing C (2014) Construction industry productivity and the potential for collaborative practice. Int J Project Manage 32(2):315–326
2. Franz B, Leicht R, Molenaar K, Messner J (2017) Impact of team integration and group cohesion on project delivery performance. J Constr Eng Manag 143(1)
3. Barbosa F, Woetzel J, Mischke J, Ribeirinho MJ, Sridhar M, Parsons M, Bertram N, Brown S (2017) Reinventing construction: a route to higher productivity. McKnsey & Company, New York
4. Hall DM, Algiers A, Levitt RE (2018) Identifying the role of supply chain integration practices in the adoption of systemic innovations. J Manag Eng 34(6):04018030
5. Noktehdan M, Shahbazpour M, Wilkinson S (2015) Driving innovative thinking in the New Zealand construction industry. Buildings 5(2):297–309
6. Hall DM (2018) Cracks in the mirror: conceptualizing the ongoing AEC industry re-organization. In: Proceedings of the 16th Engineering Project Organization Conference (EPOC 2018), Brijuni, Croatia, pp 458–477

7. Cleves JA, Dal Gallo L (2012) Integrated project delivery: the game changer. American Bar Association Meeting: Advanced Project Delivery. American Bar Association: Forum on the Construction Industry, Las Vegas, NV
8. Park M, Na LJ, Ofori G (2006) Stimulating construction innovation in Singapore through the national system of innovation. J Constr Eng Manag 132(10):1069–1082
9. Lahdenperä P (2012) Making sense of the multi-party contractual arrangements of project partnering, project alliancing and integrated project delivery. Constr Manag Econ 30(1):57–79
10. Cao D, Li H, Wang G, Luo X, Yang X, Tan D (2017) Dynamics of project-based collaborative networks for BIM implementation: analysis based on stochastic actor-oriented models. J Manag Eng 33(3):1–12
11. Fischer M, Ashcraft H, Reed D, Khanzode A (2017) Visualizing and simulating building performance. Integrating project delivery. John Wiley & Sons, Hoboken, NJ, USA, pp 265–294
12. Whyte J (2019) How digital information transforms project delivery models. Project Manag J 875697281882330
13. Lavikka R, Kallio J, Casey T, Airaksinen M (2018) Digital disruption of the AEC industry: technology—oriented scenarios for possible future development paths. Constr Manag Econ 36(11):635–650
14. Duhon B (1998) It's all in our heads. Inform 12(8)
15. Hegazy FM, Ghorab KE (2014) The influence of knowledge management on organizational business processes' and employees' benefits. Int J Bus Soc Sci 5(1):148–172
16. Hanifan G, Sharma A, Newberry C (2014) The digital supply network: a new paradigm for sup-ply chain management. Accenture Strategy
17. Duke P, Higgs S, McMahon WR (2010) Integrated project delivery: the value proposition; an owner's guide for launching a healthcare capital project via IPD. KLMK Group, Richmond, VA
18. Hall DM, Scott WR (2019) Early stages in the institutionalization of integrated project delivery. Project Manag J 50(2):875697281881991
19. Cheng R, Allison M, Sturts-Dossick C, Monson C, Staub-French S, Poirier E (2016) Motivation and means: how and why IPD and lean lead to success. Lean Construction Institute (www.leanconstruction.org)
20. El Asmar M, Hanna A, Loh W (2016) Evaluating integrated project delivery using the project quarterback rating. J Constr Eng Manag 142(1)
21. El Asmar M, Hanna AS, Loh W (2013) Quantifying performance for the integrated project delivery system as compared to established delivery systems. J Constr Eng Manag 139(11):1–14
22. Mollaoglu-Korkmaz S, Swarup L, Riley D (2013) Delivering sustainable, high-performance buildings: influence of project delivery methods on integration and project outcomes. J Manag Eng 29(1):71–78
23. Mesa HA, Molenaar KR, Alarcón LF (2016) Exploring performance of the integrated project delivery process on complex building projects. Int J Project Manage 34(7):1089–1101
24. McGraw Hill (2014) Smart Market Report The business value of BIM for owners. McGraw Hill Construction, Bedford, MA
25. Al Ahbabi M, Alshawi M (2015) BIM for client organisations: a continuous improvement approach. Constr Innov 15(4):402–408
26. Picon A (2016) From authorship to ownership. Archit Des 86(5):36–41
27. Talamo C, Atta N, Martani C, Paganin G (2016) The integration of physical and digital urban infrastructures: the role of "big data". TECHNE J Technol Archit Environ 11(11):217–225
28. Peltomaki A (2008) Living labs for user-driven open innovation, an overview of the living labs methodology, activities and achievements. Office for Official Publications of the European Communities, Luxembourg
29. Basile C, Bonanomi M, Corazza F, Manenti F, Moretti R, Pollio A, Rampanti L (2013) DeCoSe. Designing collaborative services to boost social innovation in smart cities. In Multidisciplinarity and innovation—ASP projects 7, Polipress (ed.), Arti Grafiche Fiorin, Sesto Ulteriano (MI), Italy, pp 30–39

Conclusions

This collection of essays on the theme of circular economy, prevention and enhancement of scraps and pre-consumer waste provides a framework on the policies, actions and trends that can be observed in Europe. The survey carried out on the actions promoted within the EU aims to introduce the process that led to the current systems of orientation and management of the environmental issue. From the study, it emerges that a lot has been done and is still being done on this subject. Given the success of the application of environmental programs and community financing systems, the interest and the support for this kind of initiatives are increasingly growing. In this regard, no trend reversals have ever been found, on the contrary, more and more has been done, indication that the environmental issue is a topic of primary interest in the current European scenario.

An aspect that demonstrates the interest to transform and innovate the actual system of producing and consuming derives from the study conducted on projects that have obtained European funding in the field of sustainable innovation, in particular the projects financed within the Life, CIP and Horizon programs. The study conducted highlights some fundamental aspects:

- First, over the years—both in financing and subsequent refinancing systems—there has never been a reduction of funds, on the contrary, there is an exponential growth both of the amount of funds available and of the number of funded projects.
- Second, through these financing systems different types of productive realities have been favored, above all small and medium-sized enterprises which could hardly have started strategic and radical production innovation plans without adequate financial support. Furthermore, participation in community programs has led to the activation of industrial synergies and partnerships between actors with similar interests and, therefore, it supported the creation of efficient and smart industrial networks.
- Third, the heterogeneity of the financed projects. Indeed, almost all the commercial sectors (referring to the NACE list) have received European contributions for the innovation. This aspect shows that all the productive areas have understood the

M. Migliore et al., *Strategies for Circular Economy and Cross-sectoral Exchanges for Sustainable Building Products*, Springer Tracts in Civil Engineering,
https://doi.org/10.1007/978-3-030-30318-1

need of innovation with a view to sustainability. Furthermore, the focus on the construction sector has shown that there is a high interest in innovation processes that involve the valorization of waste as a second raw material through upcycling procedures and not just downcycling.

The study conducted on these projects laid the basis for the Ri-Scarto research project presented in Chap. 5, which, having recognized the importance of sharing good practices (initiated or already consolidated in other production contexts), studied and created a prototype of a multi-sectoral platform, that could allow to share resources and information among all the actors who can make strategic choices in the industrial field, favoring the creation of virtuous networks. This last aspect has been studied in more detail because the definition of the operational limit and the limit of effectiveness of networks between companies can represent a barrier, especially if a proper analysis for defining and quantifying the avoided/induced environmental impacts, that can justify specific actions, is not carried out. One of the objective of the research was the comprehension of the relationships that can be established between the different actors involved and what are the tools and activities that can be used in order to activate circular economy scenarios. The result that emerges is precisely the need to standardize the information that can connect the different actors. This factor can allow a rapid traceability of: the possible actions that can be implemented, the possible actors that can be involved in synergic activities and the finding of materials (useless for the producers and very useful for the seeker).

In this regard, the different types of product conformity certificates were presented. Highlighting that, although they are very useful in for quantifying and qualifying products from an environmental point of view, they are still quite heterogeneous from an informative point of view. Undoubtedly, these attestations are useful both for the comprehension of the nature of the products, and for qualify them under the environmental profile (orienting the choices of the decision makers), but, it is necessary an effort to be able to make them more homogeneous. The goal is to bring out the added value deriving from a sustainable approach in the production phase, highlighting the sustainable characteristics of the product. An interesting example to justify the necessity and usefulness of the attestations of conformity is the wish to declare the effective compositions of the product and, in particular, the content of recycled material. This factor highlights the level of sustainability (percentage of recycled content) and also the possible connections that can be activated between different sector (intersectionality of the recycling chain). Currently there are more and more frequent cases in which very different productive sectors are able to find a connection in the exchange of resources, as in the case of agriculture and construction. The case study presented in Chap. 6 highlights how a problem for a sector (waste and impacts deriving from their disposal) can become a valuable resource for a radically different sector.

Lastly, the new frontiers of waste management were presented, starting from the topic of advanced waste management enabled by the adoption of innovations in the field of Information and Communication Technologies (ICTs). Nowadays, it is fundamental to recognize the role of technological innovations in the ICT field as useful support tools for establishing an active and collaborative inter-sectorial network aimed at identifying, assessing and enabling industrial symbiosis processes. Indeed, the complex nature of the waste management challenge requires coordination, collaboration and pooling of resources for defining global goals and strategies. In order to achieve these objectives, the concept of information platform is introduced as a potential interoperable framework aimed at sharing, integrating and exchanging information between stakeholders at different territorial scales. The approaches to this topic by the United Nations Environment Program (UNEP) and the European Union (EU) are also introduced and discussed, along with virtuous Smart Waste Management projects funded by the European Commission under the Horizon 2020 programme.

Finally, the adoption of new forms of integrated design and integrated project delivery models has been presented as a strategy to fostering more sustainable processes and products in the building sector. To achieve this, the shift towards collaborative and cross-functional partnerships between project participants is proposed as a driver of innovation adoption. In addition to more integration between disciplines and across phases in the building process, the participation of all the parties involved in the process is also proposed as enabling factor for boosting collaborative and sustainable services, such as the Ri-Scarto (Chap. 5) and &CO (Chap. 9) projects. Therefore, on one hand, the building process should become more integrated in order to enable project stakeholders to input their specific knowledge since the early phases of the process, taking advantage also of the new digital technologies and processes, such as Building Information Modeling (BIM). Additionally, project delivery models and management processes should change accordingly in order to facilitate the creation of cross-functional networks, which can help to achieve a better sustainability performance of both building and urban systems. On the other hand, participation of all the parties, from owners to end-users, should be fostered through innovative approaches, such as the Living Lab method, in order to achieve socio-technological innovations, which can help to give solutions and make meaningful steps towards a more sustainable society.

The manufacturer's authorised representative in the EU is Springer Nature Customer Service Centre GmbH, Europaplatz 3, 69115 Heidelberg, Germany. If you have any concerns regarding our products, please contact ProductSafety@springernature.com

Printed and bound by CPI Group (UK) Ltd, Croydon, CR0 4YY
30/09/2026
02225450-0014